FAZALE RANA

# DINOSAUR BLOOD AND THE AGE OF THE EARTH

rtb PRESS

Published by RTB Press
818 S. Oak Park Rd.
Covina, CA 91724
www.reasons.org



Cover and illustration design by 789, Inc.
Interior design by Christine Talley, Grüv Creative Communications



Library of Congress Cataloging-in-Publication Data

Rana, Fazale, 1963- author.
Dinosaur blood and the age of
the Earth / by Fazale Rana.
pages cm
Includes bibliographical references and index.
ISBN 978-1-886653-18-4

1. Earth (Planet)--Age--Religious aspects--
Christianity. 2. Dinosaurs. 3. Paleontology.
4. Religion and science. I. Title. II. Title: Dinosaur
Blood and the Age of the Earth.

QE508.R36 2016 551.7›01
QBI16-600014

ISBN 978-1-886653-18-4

Printed in the United States

# Contents

# List of Figures

# Dedication

To Hugh Ross

Your tireless work in apologetics and evangelism serves as an inspiration to me and countless others.

# Acknowledgments

Believe it or not, there were many people who worked hard and sacrificed to make this book possible. First, I want to thank my wife, Amy, for her love, encouragement, and patience when this book project took "priority" over family matters.

I especially want to acknowledge the editorial team at Reasons to Believe who dedicated themselves to this book as if it were their own. Thank you Joe Aguirre, Sandra Dimas, Maureen Moser, and Amanda Warner for your expert editorial guidance. Your efforts transformed my draft manuscript into a book that I am proud to call my own. Thanks to Charley Bell and Richard Silva from 789, Inc. for designing the book cover and the interior artwork. Also thanks to Christine Talley for designing the interior layout of the book, and Sean Platt for designing a key figure.

I'm indebted to the scholar team at Reasons to Believe. Thanks to Hugh Ross, David Rogstad, Jeffrey Zweerink, AJ Roberts, and Kenneth Samples for many insightful conversations in the hallway and during lunch. These discussions helped to directly and indirectly shape the contents of this book. Thanks to Hugh Ross, Jeff Zweerink, AJ Roberts, Ken Wolgemuth, and Richard Deem for their valuable review of the manuscript. Your input helped improve the technical accuracy and clarity of the book.

I am also grateful to each member of the Reasons to Believe team. You have supported me with your friendship and encouragement through the years. I am honored to call you friends. I am also grateful to Hugh and Kathy Ross for their friendship and guidance. Your influence has had a profound and lasting influence on my life.

INTRODUCTION

# What's the Issue?

As a kid I loved reading comic books. I purchased most of my collection at a little market store called Fannin's, just down the street from my house. Mr. Fannin sold comic books without the covers—for a penny apiece! I once asked him how he acquired his unusual inventory. He told me that the owner of the local drug store tore the fronts off comic books that didn't sell, sending the covers back to the publisher, and—horrors of horrors—threw away the denuded comic books. Fannin somehow figured out a way to retrieve the discarded comic books and sold them for next to nothing at his store. Fannin's resourcefulness was a bonanza for me.

Fond memories abound of afternoons spent scouring the campus grounds of West Virginia University Institute of Technology (where my father taught) collecting soda bottles that I would later exchange for a stack of comic books at Fannin's Market. Because the comic books cost only a penny, I wasn't all that discriminating in my selection of titles. As a result, I wound up with all sorts of comics in my coverless collection, including books published by Gold Key Comics that were part of a *Ripley's Believe It or Not!* series.

These comics were my first introduction to the *Ripley's Believe It or Not!* franchise. Founded by Robert Ripley (1890–1949), *Ripley's Believe It or Not!* began as a cartoon panel series that was first published in 1918 and syndicated in 1929 after Robert Ripley developed a readership in the *New York Globe* and later with the *New York Evening Post*. Ripley's comics focused on bizarre facts that were so unusual that it seemed questionable as to whether or not they were true. At the height of its popularity, *Ripley's Believe It or Not!* had a readership of over 80 million people. Ripley parlayed his comic panel's popularity into a radio program, and finally a TV show. His legacy lives on today in the form of a global company called Ripley Entertainment Inc. In addition to airing TV programming and publishing books of oddities, Ripley Entertainment has a number of attractions around the world, including museums (dubbed as "Odditoriums").

I became reacquainted with *Ripley's Believe It or Not!* soon after moving to Southern California with my wife and family in 1999. As newcomers to the Los Angeles area, one of the first things we did when we had a free weekend was visit Hollywood. Imagine my delight to discover a *Ripley's Believe It or Not!* museum on Hollywood Boulevard. It triggered a flood of memories for me. I convinced my wife and kids to join me that afternoon as I went through all the attractions. They weren't all that impressed, but I had a blast, surrounded by wall-to-wall oddities and exhibits commemorating remarkable people who had accomplished unbelievable feats.

Or had they?

why is it hard to accept? Because he has precondit…
The idea that the fossile is 80 million yrs old is fact

---

Robert Ripley traveled to 201 countries in 35 years seeking out oddities to feature in his comic panel. But Ripley did not have the market cornered on hard-to-believe facts. Scientists have discovered their share of remarkable finds as they have investigated the world around us. Perhaps one of the hardest-to-accept discoveries made in recent years has been the recovery of original soft tissue remnants within the fossilized remains of dinosaurs that lived nearly 80 million years ago. As paleontologists have traveled around the world and explored and carefully examined fossilized bones, they have repeatedly unearthed dinosaur specimens that have yielded soft tissue remnants. These preserved tissues include capillary-like structures, cell remnants, protein fragments, and skin and feather pigments. Plus, the discoveries are not limited to dinosaurs. Paleontologists have also uncovered: (1) quinones from sea lily fossils that date to 350 million years in age; (2) ink within the ink sacs of fossilized cuttlefish, dating to 160 million years in age; (3) chitin from 34-million-year-old cuttlefish and 505-million-year-old sponges; and (4) shell proteins from 15 million-year-old mollusk fossils. I question every one of these ages

What an unbelievable stroke of "good luck" for paleontologists. Common wisdom has long held that soft tissues should readily degrade in a few thousand years. Yet, in some cases, these biomaterials have persisted for a few hundred million years! Such unexpected discoveries excite the research community because they open up the possibility for scientists to gain important insight into the biology of ancient organisms—insight they never thought possible.

He treats this as an undeniable fact. But it isn't.

---

These discoveries also excite another group of people: Christians who believe that the earth is young. They see these breakthroughs as compelling scientific support for their interpretation of Genesis 1—one that regards the creation days as calendar days. Accordingly, the creation week spans six back-to-back, 24-hour periods and, consequently, Earth must be merely 6,000 years old. Young-earth creationists (YECs) have capitalized on these surprising results to argue that it is impossible for the fossils to be millions of years old. They reason that soft tissues should not survive that long. Instead, they should degrade in a few thousand years. Thus, the scientifically determined date for the fossils must be in error. In their view, these finds challenge the reliability of radiometric dating methods used to determine the age of these fossils and, along with it, Earth's antiquity. They contend that these breakthrough discoveries provide compelling scientific evidence for a *young* earth and support the idea that the fossil record results from a recent global (worldwide) flood.

In fact, this line of reasoning has become one of the most prominent scientific arguments for a young earth. A visit to Ripley's is not necessary to see why this is the case. Very few, if any, paleontologists thought that soft tissue remains would be preserved within fossils, particularly those that date to several hundred million years in age. Organic compounds, which comprise soft tissues, are typically rather "delicate" materials that readily break down.

Scientists and other scholars from young-earth creationist organizations, such as the Institute for Creation Research (ICR) routinely monitor the scientific literature, identifying and commenting on age-of-the-earth implications of each new discovery of fossilized soft tissue remnants. While many of their articles are written for lay readers, the YEC researchers have also produced impressive technical analyses of the field.[1] YECs have also launched their own research program in fossil soft tissues, generating original scientific findings. One of these efforts has been dubbed the iDINO (Investigation of Dinosaur Intact Natural Osteo-tissue) project. Biologist Mark Armitage identified the first evidence for soft tissue remnants in the fossilized horn of a *Triceratops* specimen that dates to around 68 million years in age.[2] In the spring of 2015, fellow YECs Brian Thomas and Vance Nelson reported the radiocarbon dates for soft tissue recovered from dinosaurs and other fossils, claiming that this data indicates these materials are only about 40,000 years in age.[3]

All of this activity has influenced many Christians to use these arguments when they engage skeptics and seekers alike with evidence for the credibility of the Christian faith. Are they correct?

As this book will demonstrate, few people in the scientific community are impressed with this latest scientific argument for a young earth. Even though the recovery of soft tissue from fossil remains was unexpected,

Well of couse they are not impressed. their preconceptions dictate that the EARTH is old. They will naturally scoff at any evidence to the contrary the soft tissue discoveries trouble virtually no scientists. It would be challenging to find an expert who questions the measured ages of fossils and the antiquity of the earth because of the recovery of soft tissue remnants from fossils. In fact, most in the scientific community look askance at such young-earth arguments. of course they're brainwashed

Joining these scientists are Christians who accept the scientific ages for the fossil record and Earth's age. These Christians (including me) have great concerns about the impact that young-earth arguments can have on evangelism. Many skeptics and seekers alike reject Christian truth claims because they wrongly think Genesis 1 teaches that the earth is only 6,000–10,000 years old. This misperception is reinforced by vocal YECs who not only claim that the only valid interpretation of Genesis 1 is the calendar-day view (six 24-hour days), but also maintain that ample scientific evidence—such as recovery of soft tissue remnants in fossils—exists for a young earth. But these questionable scientific claims can cause many people to view Christianity as antiscientific, undermining outreach efforts.

Tragically, nonbelievers may be hindered from engaging the powerful scientific evidence for God's existence and Scripture's reliability. This leads us to the purpose of this book; namely, to help Christians understand why it makes sense—from a biochemist's standpoint—for soft tissue remains to be preserved in fossils that date to several hundred million years in age. As a biochemist, I understand the structure, function, and *stability* of molecules (such as those found in soft tissue remains). I hope that my insights can both help prevent well-meaning believers from making a scientifically questionable argument for a young earth, and can also help Christians in general to exercise discernment when evaluating arguments for the credibility of the Christian faith.

---

The book is organized into four main chapters. The first will describe the dramatic story of the discovery of soft tissue remnants in a *Tyrannosaurus rex* and also summarize some of the other key soft tissue finds. In the next chapter, I will do my best to fairly represent the reasons why YECs think these discoveries serve as positive evidence for a young earth.

As I have pointed out, YECs insist that the scientifically determined dates for the fossils must be in error, because of the soft tissue recovered from these remains. In their view, these findings challenge the reliability of radiometric dating methods used to determine the age of these fossils, and along with it Earth's antiquity. The goal of chapter three is to demonstrate why radiometric dating *is* trustworthy. Rather than a comprehensive

why doesn't he think these dates could be in error?

discussion of the techniques, I will provide a brief description of how scientists use these methods to date fossils, referring interested readers to works that give a more detailed discussion of radiometric techniques. I will then offer a few reasons why I think these methods produce believable results.

If radiometric dating methods *are* reliable, then how is it possible for soft tissue remnants to persist for millions of years? Chapter four addresses this question. Though unexpected, the presence of soft tissue remnants in fossils can be accounted for if the burial and fossilization process occurred under specific sets of conditions. In addition, researchers have also discovered additional mechanisms that can stabilize soft tissues, ensuring their survival for millions of years. This chapter will also discuss different structural features of certain biomolecules and soft tissue types that promote their long-term persistence.

I have tried my best to make the science accessible to lay readers and to avoid technical jargon, but in many instances it simply was not possible. I have put together a short glossary to help readers understand some of those terms. I would also encourage nonspecialists to focus on the "big picture" ideas when they feel overwhelmed with the terminology.

A secondary goal for this work is to help Christians overcome unnecessary obstacles to old-earth creationism. I hope to accomplish this task by including an appendix that touches on the biblical case for an old earth. Two other appendices follow, describing some other hard-to-believe discoveries involving dinosaurs that have important apologetics implications.

Robert Ripley asserted that he could prove every hard-to-believe claim he made. He exuded that confidence because he employed a professional fact checker named Norbert Pearlroth, who not only dug up many of the oddities that were the subject of Ripley's comic strips, but also checked the claims submitted by readers.

For some readers, it will be hard to believe that soft tissue remnants associated with fossil remains could be tens or even hundreds of millions of years old. But I hope as you check out the facts for yourself that you will come to believe it is true. you see the bottom line is that it's all about beliefs, not facts! facts have to be interpreted.

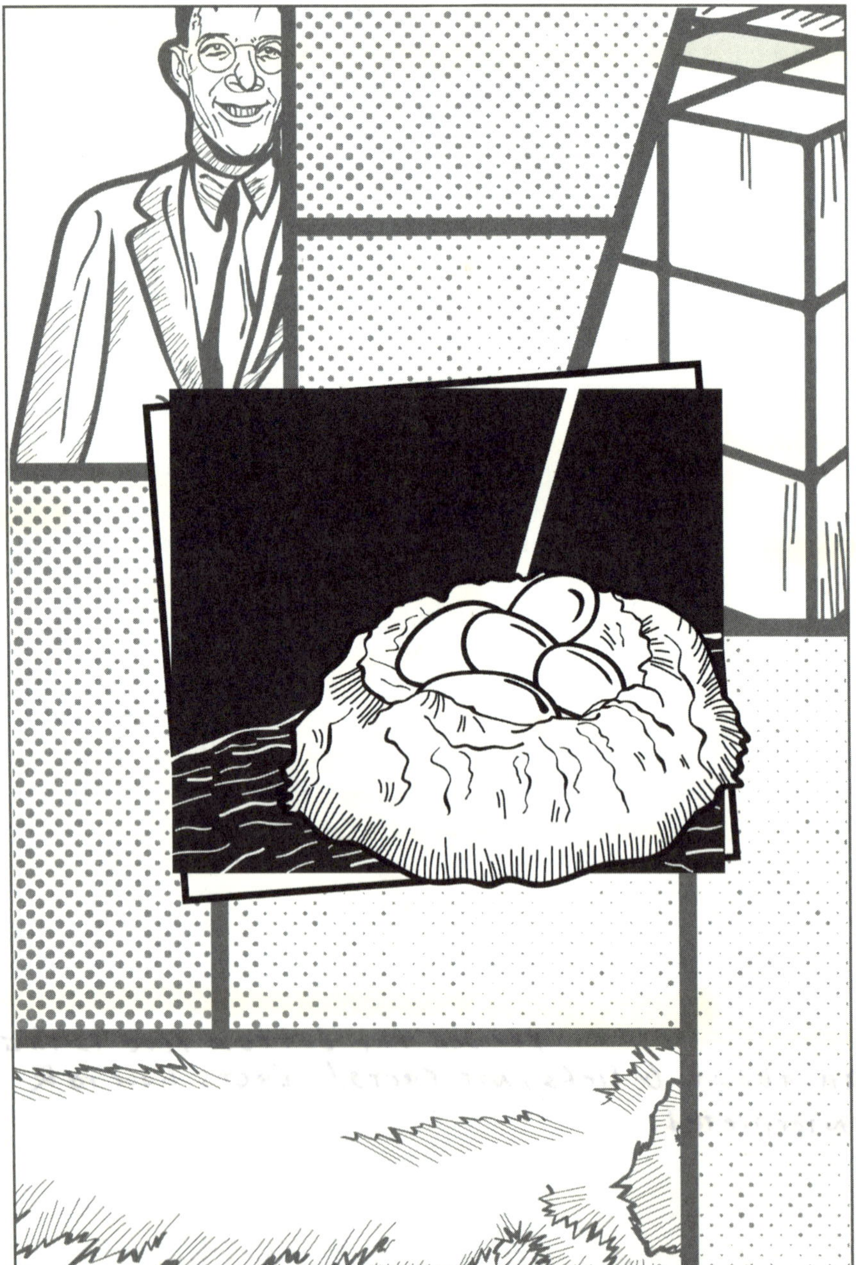

CHAPTER 1

# Dinosaur Blood in Fossils: Who Would Believe It?

One of my favorite displays at *Ripley's Believe It or Not!* Odditorium in Hollywood shows the replicas of 100-million-year-old dinosaur eggs. Scientists discovered the originals in China in the 1920s. They were found clustered together as if they were deposited in "nests." Paleontologists believe that a *Hadrosaurs* laid the ancient eggs.

These Chinese fossils are only one among many such treasure troves. Since their discovery in the 1920s, fossilized eggs have been unearthed all over the world. Believe it or not, paleontologists have recovered dinosaur eggs that are, in some instances, nearly 200 million years old! Even more remarkable is the fact that scientists, using CAT scan technology and X-ray imaging techniques, can detect dinosaur embryos within some of the better-preserved specimens.

Perhaps even harder to accept is the claim by scientists from Germany (in 2015) that certain dinosaurs laid blue-green eggs.[1] Researchers reached this conclusion after detecting protoporphyrin and biliverdin in three separate sets of dinosaur eggs (dating to 66 million years in age) recovered in China. They think that these eggs were laid by the oviraptor *Heyuannia huangi* in open-air nests.

Protoporphyrin and biliverdin are responsible for the brown color of chicken eggs, the blue color of a robin's eggs, and the blue-green coloring of emu eggs. The different colors result when these compounds are blended together in different proportions. Specific interactions between these pigments, the proteins, and the calcium carbonate that makes up the eggshells also contribute to the egg's hue.

The relative proportion of the two compounds in the dinosaur eggs is similar to that found in the blue-green eggs of emus, leading the scientists to speculate that these dinosaur eggs were probably blue-green, as well. This interpretation gains support from the fact that the dinosaur eggs were found in open nests. Emus lay eggs in these types of nests also. The blue-green coloring provides camouflage for the emu eggs, making them more difficult for predators to spot than white eggs. Presumably, the same was true for the *Heyuannia huangi* eggs.

This remarkable insight is possible because organic molecules survived in the fossilized eggshells—for 66 million years! A few years ago, no one would have thought that organic molecules could survive long enough to be recovered from fossilized eggs. No one in their right mind would have been willing to spend time, effort, and money to try to detect organics in fossilized specimens as the German scientists had done in 2015. Yet, that unwillingness changed in 2005 after a team of researchers, headed by paleontologist Mary Schweitzer, detected organic materials for the first time in fossilized eggshells.[2] That pioneering work was a scientific tour de force.

Schweitzer and her team studied the shells of fossilized eggs that were about 70 million years old, discovered in Argentina. The shells surrounded dinosaur eggs housing unusually well-preserved embryos, including fossilized embryonic dinosaur skin. Paleontologists reasoned that the fossilization process for the specimens was so rapid that it occurred before the organic materials completely degraded. On this basis, Schweitzer and her team thought there was a chance that organic materials may have survived in the eggshells. To test this idea, they ground up pieces of the eggshell fossils and extracted them with the hope of isolating any organic material that survived. To everyone's surprise, they were able to detect fragments of the egg protein ovalbumin.

The success of Schweitzer's work no doubt motivated the German scientists to probe for the presence of pigments in ancient oviraptor eggs. But if the common wisdom at that time held that it was impossible for organic materials to survive for hundreds of thousands—let alone millions—of years, what inspired Mary Schweitzer to look for organic materials in dinosaur eggshells? Schweitzer's motivation came from her serendipitous discovery in 1992, when she was a graduate student at Montana State University working with legendary dinosaur paleontologist Jack Horner.[3] This discovery would transform paleontology, opening up new vistas no one could have imagined.

---

For those conversant with dinosaur paleontology, the story of Schweitzer's discovery of soft tissue remnants in dinosaur bone is well-known.[4] It is worth briefly recounting the story (for those unfamiliar), because it wonderfully illustrates how much of a role serendipity sometimes plays in science and how having the courage to take calculated risks can lead to huge payoffs.

Mary Schweitzer was a nontraditional graduate student. Instead of enrolling in graduate school immediately after earning her undergraduate degree, Schweitzer started her graduate studies in paleontology at Montana State University later in life. She had started a family when she went back to school to earn a degree in science education. Schweitzer put her degree to use, taking a position as a substitute teacher. But she missed the intellectual stimulation of the college environment, and decided to audit a class in paleontology from Jack Horner just for fun. Because she was so enamored with dinosaur paleontology, she went on to serve as a volunteer in Horner's lab at the Museum of the Rockies after completing the class. During her time as a volunteer, Schweitzer became notorious for asking questions, to the point that Horner finally advised her to go to graduate school.

Schweitzer's research project in graduate school seemed to be relatively "safe and straightforward." She planned on using state-of-the-art imaging techniques to study the microanatomy of dinosaur bones, looking for differences in load-bearing and nonload-bearing bones. Having difficulty preparing thin-enough slices of the fossilized bones so they could be studied, Schweitzer asked for help from a friend in the veterinary department who happened to be an expert in bone tissue.

Things took a dramatic turn a short time later when her friend was presenting a talk at a scientific conference and she was asked what was the oldest bone she had ever worked on. In response, the researcher pulled out a slide she had prepared of a thin section of Schweitzer's fossilized dinosaur bone and projected it so the audience could see. Afterward, a scientist who attended her talk came up to her and pointed out that there were red blood cells present in the slide of the fossilized bone of the *T. rex* femur—that was 68 million years old!

Schweitzer hesitated to let Horner know what she and her friend discovered—at least until she could get her scientific "ducks in a row." After all, to have a graduate student claiming that soft tissues may have survived over 68 million years could have brought major embarrassment to a scientist of the stature of Jack Horner. However, during that process, word made it to Horner about Schweitzer's activities. He wasn't happy that she was pursuing her own research agenda, gallivanting around the Montana State campus with "his" dinosaur fossils. But his anger quickly subsided, replaced by stunned silence and then scientific curiosity about what Schweitzer had stumbled upon. Within the bone channels, where blood vessels would have been when this *T. rex* was alive, were ruby-colored structures that were the right size, shape, and color to be red blood cells. Also, it looked like these "cells" possessed nuclei. This observation was significant. While mammalian red blood cells lack nuclei, those of birds and reptiles possess these subcellular structures. It made sense that dinosaur red blood

This idea that the earth is incredibly old is so strong that it clouds his thinking. He is incapable of truly engaging in critical thinking about this subject cells should harbor these organelles as well.

Horner challenged Schweitzer to prove that the structures in the bone *were not* red blood cells. Suddenly Schweitzer's project changed from a relatively safe endeavor to one that would be extremely controversial and risky for an experienced, well-established scientist to pursue, let alone a graduate student who was a scientific neophyte. But Schweitzer boldly took up Horner's challenge. She tried her best to prove that the red microstructures in the dinosaur bone were not red blood cells—and she is still trying today.

---

In 1997, Schweitzer, Horner, and other collaborators published two papers that summarized their first attempts to prove that the objects in the bone *were not* the remnants of red blood cells.[5] Yet, they failed in that regard. The evidence they generated seemed to confirm their initial suspicions. Schweitzer and her collaborators detected amino acids (the building blocks of protein) in extracts of the fossilized *T. rex* bone—including glycine and hydroxylysine, two of the most prominent amino acids in bone collagen. The amino acids recovered from the bone extract were a mixture of L- and D- varieties, as opposed to exclusively the L- form. This indicates that the amino acids did not result from contamination, but had been in the bone for a significant period of time—long enough to undergo a chemical process called racemization. The presence of collagen fragments in the bone means that it is reasonable to think that the ruby microstructures were actually red blood cell remnants. If collagen remnants could persist in the fossils, so could remnants of red blood cells.

Perhaps the most impressive result the team obtained came when they observed antibodies, generated from fossil bone extracts, binding to hemoglobins from turkeys and rats. (Hemoglobin is the protein that plays a role in oxygen transport and it occurs in high concentrations in red blood cells.) To generate these antibodies, Schweitzer and her collaborators injected the fossil bone extracts into rabbits. The rabbit's immune system treated the extracts as a foreign material, producing antibodies against it. Antibodies are a special class of proteins that take part in the immune response. When foreign materials and pathogens enter the body, antibodies bind to these harmful entities, targeting them for attack by the immune system. Antibody binding can be *highly* specific. The fact that antibodies raised against fossil bone extract bound turkey and rat hemoglobins is a strong indicator that hemoglobin, or at least hemoglobin fragments, are present in the bone extract. Using two different spectroscopic techniques, Schweitzer and her team also detected evidence for heme in the fossilized bone extracts. Heme is the portion of the hemoglobin molecule that binds oxygen. Heme's presence in bone makes sense if the red microstructures were indeed remnants of red blood cells.

When Schweitzer reported these finds, she was very cautious about the conclusions she drew regarding the two studies, carefully noting that the results were merely *consistent* with the presence of red blood cell and protein remnants (soft tissues) in the dinosaur bones. Because Schweitzer and her collaborators were so tentative about their conclusion and so conservative about their interpretation, their work "got very little notice," according to Schweitzer.[6]

But a follow-up study was about to change that, propelling Schweitzer and her colleagues into the limelight.

In the summer of 2003, an excavation crew finished unearthing a remarkable *T. rex* specimen from Hell Creek Formation in Montana, dating to 68 million years in age. During the process of transporting the massive skeleton from the site, small bone fragments broke off one of the femurs. Schweitzer and her team got a hold of the pristine fossil samples with the hope of finding more evidence for cellular and molecular materials in dinosaur bone.

When Schweitzer's team dissolved away the mineral matrix using a weak acid and chelating agent (ethylenediaminetetraacetic acid), they were shocked to find the structural remains of blood vessels that were flexible, transparent, and hollow.[7] Inside the blood vessel remnants were red, round microstructures that came out of the vessel's interiors into solution. These microstructures displayed cell-like morphology. Flexible and resilient fibrous-like structures were also left behind after demineralization. Schweitzer and her team demonstrated that the flexible, fibrous-like components contained collagen fragments and even determined the amino acid sequences of those fragments, confirming their identity.[8]

Schweitzer and her team had made a discovery worthy of its own display at a *Ripley's Believe It or Not!* Odditorium: *T. rex* blood vessels! If that weren't enough, the team even found structures that looked like osteocytes (with filopodia) floating around in solution after the demineraliztion procedure. Osteocytes produce collagen and other materials that make up the organic portion of bone. To confirm that these cell-like structures were indeed osteocyte remnants, Schweitzer's team showed that fragments of proteins (such as actin, tubulin, PHEX and histone 4) that are associated with osteocytes from extant organisms are also associated with the osteocyte-like structures recovered from the ancient *T. rex* specimen.[9] These investigators even claimed to have observed antibodies designed to bind to DNA binding to the osteocyte-like structures, raising the possibility that pieces of dinosaur DNA survived for 68 million years in the *T. rex* femur!

---

These incredible discoveries evoke the question: How prevalent are soft tissue remnants in dinosaur fossils? Do they occur widely or are they limited to only a few specimens?

In an attempt to address these questions, Schweitzer and her collaborators analyzed fossilized bone samples from another dinosaur (the hadrosaur *Brachylophosaurus canadensis*) that appeared to be well-preserved. Dating to 80 million years in age, this specimen was unearthed in Montana.[10] As with the *T. rex* specimen, demineralization of the bone yielded a flexible, transparent, hollow material that appeared to be the leftovers of blood vessels containing red microstructures. Spectroscopic analysis indicated that the blood vessel remnants were made out of a protein-like substance. Additionally, the researchers uncovered fibrous, flexible matrix materials and osteocyte-like structures. They observed the bone extracts binding to antibodies raised for collagen, osteocalcin, elastin, laminin, and hemoglobin. Amino acids (glycine, lysine, hydroxylysine, proline, hydroxyproline, alanine, and leucine) most abundant in collagen were also found in the extract. Finally, Schweitzer and her collaborators generated amino acid sequences of collagen fragments, confirming their identity.

In 2013 an international team of investigators "one-upped" Schweitzer and her team when they detected protein vestiges in the bones of dinosaur embryos, *in situ* (in place).[11] These embryos were part of a fossilized bone bed discovered near Kunming, China, in geological layers corresponding to the early part of the Jurassic. Included in this bone bed were eggshells and pieces of embryo skeletons representing different stages of embryonic growth. The scientists couldn't believe their luck. Finds such as this are extremely rare. Because of Schweitzer's success detecting collagen in extracts of fossilized dinosaur bones, this team wanted to see if they could detect protein remnants within the bone itself, forgoing the extraction process. To accomplish this task, the team used a Fourier-Transform Infrared Microspectrometer, which employs synchrotron radiation. This instrument recorded spectra corresponding to microscopic images of bone microstructures. The researchers detected signals indicative of protein decay products in fast-growing areas of embryo bone and in the bone channels that housed blood vessels.

What does it all mean?

These two additional studies, as well as a survey of dinosaur specimens conducted by Schweitzer and her team, indicate that soft tissue remnants may well be a common feature of dinosaur fossil remains.[12] Analysis of extracts from the fossilized bones of nine different dinosaur specimens, ranging in age from 67 to 100 million years and recovered from sites primarily in Montana, indicate that blood vessel vestiges, fibrous matrices, and osteocyte and red blood cell remnants are commonplace. They discovered that soft tissue remnants are distributed to varying degrees among the dinosaur specimens and differ greatly in preservation. But at least *some* soft tissue remains can be recovered in every one of the specimens.

Schweitzer's fossilized bones are well-preserved specimens, which stacks the deck in favor of soft tissue survival. A study published in 2015 by researchers from the United Kingdom suggests that soft tissue remnants may even be prevalent in poorly preserved dinosaur bones.[13] That group analyzed eight fossilized dinosaur bone fragments from the Natural History Museum, London. In one fragment they found evidence for collagen fragments, and in others, oval structures with similarities to red blood cells.

---

Fossilized bones are not the only dinosaur remains to yield soft tissue remnants. As mentioned, researchers have isolated ovalbumin fragments and pigments from dinosaur eggshells. Young-earth creationists Mark Armitage and Kevin Anderson uncovered evidence for soft tissue remnants in a *Triceratops* fossilized horn. Armitage unearthed this specimen at the Hell Creek Formation in Montana along with *Triceratops* ribs and vertebrae, discovered within a mile of the horn. Soaking pieces of the horn in a mild acid bath for a month released soft, flexible brown sheets about 8 inches by 4 inches in size. Detailed microscopic analysis of the sheets revealed the presence of osteocyte-like structures replete with delicate filipodia and nuclear-like structures within the "cells."[14]

Researchers have also detected soft tissue materials in fossilized feathers of dinosaur-like birds. Schweitzer and her group found evidence for beta-keratin in fibrous-like material associated with feather-like structures of a well-preserved specimen of *Shuvuuia deserti.* [15] The team also detected evidence of keratin fragments in fossilized claw structures of *Rahonavis ostromi.*[16] In 2015 an international team of researchers found evidence for subcellular structures called melanosomes in the fossilized feathers of the dinosaur-like bird, *Anchiornis huxleyi.*[17] This specimen was recovered in China and is dated to 150 million years in age. Melanosomes harbor the pigment melanin. Using spectroscopic methods that allowed them to gain chemical information, *in situ*, on the melanosome-like structures, the researchers concluded that these microscopic structures must be melanosomes because they contained materials that gave the same type of spectroscopic signature as animal eumelanin (one of the most commonly occurring types of melanin).

One of the most fascinating soft-tissue finds associated with dinosaur fossils (at least if you are a junior-high boy or a middle-aged biochemist who has never quite grown up) comes from the work of Karen Chin's team. (Mary Schweitzer was a member of this team, by the way.) These investigators discovered undigested muscle tissue in 70- to 80-million-year-old coprolites.[18] In case you don't know, coprolite is the term paleontologists use to refer to fossilized dung. This specimen was "deposited" by a *T. rex*. The coprolite was about two feet long and half a foot wide. The research team estimated that, when fresh, the coprolite corresponded to about 3 gallons of "material."

Amazing and disgusting! Forget the replicas of 100-million-year-old dinosaur eggs. I want to see replicas of *T. rex* coprolites in the *Ripley's Believe It or Not!* Odditorium.

Perhaps most amazing of all is the fact that the researchers were able to detect evidence of muscle cells and muscle fibers from undigested food in the coprolites, marked by characteristic banding patterns. This discovery gives researchers an unprecedented opportunity to "get the straight poop" on the feeding behavior of these remarkable dinosaurs. Apparently, these creatures gulped their food.

---

Dinosaur paleontologists are not the only ones who get to have all the fun. The pioneering work of Schweitzer and others has inspired more fossil hunters to look for soft tissue remnants in the remains of fossilized organisms. What follows is a sampling of discoveries, selected to illustrate just how pervasive soft tissue remains seem to be throughout the fossil record.

In 2015 researchers from the Carnegie Institution for Science in Washington, DC, reported that they were able to release thin protein sheets from reddish-brown fossilized shells of the gastropod *Ecphora*.[19] The researchers treated the shells with diluted acid. This treatment dissolved the mineral component (calcium carbonate) of the shell, leaving behind a flexible film. By analyzing the film, investigators discovered that it is composed of intact proteins with an amino acid composition matching that of the modern shell-binding proteins of mollusks. (The architecture of mollusk shells consists of alternating organic and inorganic layers. The organic layer is made up of sheets of shell-binding proteins and polysaccharides. The organic layer provides a surface for the deposition of minerals made up of calcium carbonate, such as calcite and aragonite.) The fossilized shells were recovered from Calvert Cliffs, a formation in Maryland that spans between 8 and 18 million years in age. Researchers estimate the age of the specimens they analyzed at around 15 million years in age. Wow!

An international team of investigators discovered evidence for heme (the oxygen-binding portion of the hemoglobin molecule) in a fossilized mosquito, dating to 46 million years in age, recovered in Montana.[20] This specimen appeared to have been engorged with blood at the time of its death. Researchers were able to detect iron within its fossilized abdomen, along with a material containing porphyrin rings. These rings make up part of the heme molecule, serving to bind the iron associated with hemoglobin.

In 2011 researchers from the UK and the US produced a detailed spectral map of 50-million-year old fossilized reptile skin.[21] This specimen was unearthed at the Green River Formation in Utah. Their analysis provided evidence that keratin remnants survived in this specimen. Keratin is a major protein component of skin.

In 2012 researchers recovered the fossilized remains of cephalopods (a class of marine animals, like squid, that squirt ink) in the UK found in rock formations that date to about 160 million years. To their surprise, they discovered that large, black ink sacs have been preserved within these specimens. Careful examination of the ink sac contents reveals what appears to be ink composed of granules similar in size and shape to those that comprise the ink of modern cephalopods. Additional characterization of the substance reveals chemical properties similar to the ink found in modern cephalopods.[22] Cephalopod ink is composed of a compound called eumelanin, a type of melanin that is dark brown-black. Another form of melanin, pheomelanin, is orange-red in appearance. (Melanin is the pigment that gives human skin, hair, and eyes their color.) These two variants of melanin play numerous biological roles, including photoprotection, radioprotection, display, camouflage, and predator avoidance. The prospect that melanin can be preserved in essentially a chemically intact form for at least 160 million years is nothing short of remarkable.

In 2013 a team of researchers from Ohio State University reported the recovery of quinones (organic compounds) from sea lily fossils that date to 340 million years.[23] Sea lilies (animals with a plantlike appearance) belong to a biological group called the crinoids. These organisms use quinones as pigments and toxins.

Researchers have also recovered chitin from the fossilized remains of several different organisms. Chitin is a hard, durable organic material that makes up the exoskeleton of arthropods and the hard structures of mollusks. In 1997 researchers from the UK and Germany provided evidence for chitin remnants in 25-million-year old fossilized insects, recovered in Germany.[24] Using an analytical technique that can separate, identify, and quantify individual components in complex mixtures, the team generated a chemical fingerprint for the fossilized materials that matched that of chitin isolated from modern-day insects.

Over a decade later, another group demonstrated that chitin remnants were found in 35-million-year-old cuttlefish fossils that had been recovered from a site in Mississippi.[25] Still another research team provided evidence for a protein-chitin complex in the cuticles of fossilized scorpions that dated to 310 million years in age and in fossilized sea scorpions that dated to 417 million in age, recovered from sites in Illinois and Ontario, Canada, respectively.[26] All of these studies are significant because they demonstrate that chitin remnants will persist for both arthropods and mollusks under both terrestrial and aquatic burial conditions.

At the end of 2013, a team of European scientists discovered the oldest chitin remains to date. This team provided evidence for chitin remnants in sponge fossils that date to 505 million years in age! The fossils were unearthed in the Burgess Shale of the Canadian Rockies, and—believe it or not—represent some of the first animals on Earth![27]

When Mary Schweitzer first published her reports of soft tissue remnants recovered from dinosaur remains, her results seemed like nothing more than a scientific oddity at the time. But as more and more paleontologists use sophisticated analytical techniques to excavate fossil remains, the recovery of the remnants of biomolecules, cells, and tissues from a wide range of fossils has become commonplace. It has also caused an upheaval of a key paradigm in paleontology.

---

Prior to Schweitzer's discovery of blood cell remnants in *T. rex* fossil remains, paleontologists thought that information about the soft tissue anatomy and physiology of ancient organisms' biology was lost to time, except for rare instances in which the soft tissue left behind imprints in the rock. According to Schweitzer, "For more than 300 years, paleontologists have operated under the assumption that the information contained in fossilized bones lies strictly in the size and shape of the bones themselves. The conventional wisdom holds that when an animal dies under conditions suitable for fossilization, inert minerals from the surrounding environment eventually replace all of the organic molecules—such as those that make up cells, tissues, pigments and proteins—leaving behind bones composed entirely of mineral."[28]

But Schweitzer's work (and that of other teams of paleontologists) calls into question the validity of this paradigm. As Schweitzer points out, "Our findings challenged everything scientists thought they knew about the breakdown of cells and molecules."[29] They also indicate that the way paleontologists think fossilization takes place may be wrong.

In spite of the impressive amount of data generated through careful, painstaking investigative work, many paleontologists remain unconvinced that soft tissue leftovers have persisted in the fossilized remains of ancient organisms. It is true that some scientists find Schweitzer's work (and that of other paleontologists working in this area) compelling. Still, others want more data. Though they are skeptical, they remain open-minded, taking a "wait and see" approach. A few have advanced alternative explanations for the soft tissue. They are motivated to do so, in part, because they can't let go of the conviction that soft-tissue cannot survive for millions of years.

For example, a team of researchers from the US and Poland have claimed that the "blood vessels" recovered from the *T. rex* bones were actually bacterial biofilms.[30] But their claim is not compelling because they did not demonstrate that biofilms could grow in fossilized bones. The explanation doesn't fit with the recovery of "blood vessels" from dinosaurs fossilized under a variety of different environmental conditions. These differing conditions should impact biofilm growth, retarding it in some cases. Yet, "vessel-type" structures have been recovered from the different dinosaur specimens. The bacterial biofilm interpretation also doesn't explain why the "bacterial films" are hollow, or why they contain microstructures that look like osteocytes, equipped with filopodia.

Another research team from the Palo Alto Research Center in California re-analyzed Schweitzer's collagen results, specifically focusing on the amino acid sequence data.[31] This group agreed that Schweitzer and her collaborators did indeed discover genuine evidence for collagen in their samples. And the collagen and hemoglobin plausibly appeared to be ancient. But, they argue that Schweitzer's team detected collagen and hemoglobin contamination from soil bacteria and birds, not authentic *T. rex* collagen. While contamination is always a concern when studying ancient biomolecules that usually occur at trace levels in the sample, the scenarios these researchers propose for the source of the contamination borders on the absurd. The researchers state, "Contamination remains a tricky and possibly unresolvable issue for this particular sample. Perhaps a bird died on top of the *T. rex* excavation in the field; perhaps ostrich bone lingered in the mass spectrometry facility for a year; or perhaps avian collagen from a cosmetic or medical product found its way into the *T. rex* sample."[32]

And how does their explanation jive with the results of other research groups? Contamination does not account for the detection of collagen and other proteins reported by other investigators. Did birds die on top of those fossil specimens as well? Did those researchers also contaminate their samples with cosmetics?

In response to these challenges, Schweitzer and a team of collaborators analyzed blood vessels from a *Brachylophosaurus canadensis* specimen recovered in Montana (dating around 80 million years in age) with the explicit goal of demonstrating that the blood vessels aren't biofilms and the protein fragments they recovered aren't due to contamination.[33] To eliminate concerns about contamination, the researchers took a number of precautions. First, they prepared duplicate samples in two separate laboratories, using dedicated facilities to ready the samples for analysis. Everyone who handled the samples wore protective clothing, and the analyses were undertaken with analytical instruments that had never been exposed to proteins from birds.

Before isolating the blood vessels from the dinosaur fossils, they carefully examined their microstructure to look for any evidence that microbes had invaded the bone. They found none. They also treated the bone with a dye that would stain the bone if there were any fungal contamination. Again, they found no evidence for fungal contamination. Once they isolated the blood vessel they carefully characterized their morphology, noting that the blood vessel structures appear to be made up of continuous vessel walls of uniform thickness. If the "blood vessels" were biofilms they should display a lack of uniformity, but they don't.

Based on analysis of their amino acid sequences, the protein fragments they isolated from the *Brachylophosaurus* blood vessels appeared to come from actin, myosin, tubulin, and tropomyosin, as expected if these blood vessels were authentic. Moreover, the researchers determined that the amino acid sequences for the protein fragments displayed signatures that are diagnostic of vertebrates. This result eliminates bacterial contamination as the source of the protein remnants in the fossils and strongly supports the bioauthenticity of the blood vessels.

It seems that some scientists are so locked into the old paradigm that no amount of evidence will convince them that soft tissue remnants exist in fossil remains. Schweitzer recounts an exchange with an anonymous scientist during the peer review process for an article she submitted to a scientific journal for publication. The reviewer wrote that "this type of preservation was not possible" and that "I could not convince him or her otherwise, regardless of our data."[34]

When Schweitzer first saw evidence of red blood cells in *T. rex* bones, she knew then that controversy would surround her work. And it has. But her work has also generated much excitement.

---

While some paleontologists have been lukewarm about the survival of soft tissue remnants during fossilization, one group of Christians has warmed up to this possibility rather quickly. They don't see Schweitzer's work as controversial whatsoever. In fact, they claim that they predicted soft tissues would be found in fossils all along. These Christians are young-earth creationists (YECs), who contend that the earth and life on Earth are only 6,000–10,000 years old. YECs insist that the discovery of soft tissue vestiges in fossils demands an upheaval of a scientific paradigm. But the paradigm they refer to has nothing to do with the capability of organic materials to survive during the fossilization process. As I will discuss in the next chapter, YECs argue that these discoveries directly challenge the scientific consensus on the earth's age and the antiquity of life on Earth.

CHAPTER 2

# Dinosaur Blood and the Case for a Young Earth

Of all the odd displays in the Hollywood *Ripley's Believe It or Not!* Odditorium, perhaps the most unusual is the fur-covered trout. That's right—a fur-covered trout! This fish is covered with a big coat of white fur. Even though its authenticity is a bit "fishy," this trout was displayed at the National Museum of Scotland as a genuine Canadian species. Apparently, the museum curators adamantly refused to believe the fish was a hoax, and as a result it remained on display well into the 1930s. I don't know which is harder to believe: the existence of a fur-covered fish or the gullibility (or maybe the stubbornness) of the museum curators.

Young-earth creationists argue that the scientific community has indoctrinated a gullible public into believing an evolutionary-based dating of the earth (at 4.5 billion years old) and of the fossil record (with fossilized animal remains as old as 570 million years in age). These Christians interpret the "days" in Genesis 1 as 24 hours in duration. Consequently, they maintain that Earth and its life can only be 6,000–10,000 years old. YECs point out that many people lack a foundational scientific and biblical understanding of the evidence for a recent and direct creation. These people then become easy prey to the claims of the scientific community—claims steeped in the evolutionary paradigm (which they say includes the antiquity of Earth and past life).

According to Kevin Anderson of the Van Andel Creation Research Center and head of the Creation Research Society's iDINO (Investigation of Dinosaur Intact Natural Osteo-tissue) project, "In contemporary studies of science, evolution is simply assumed to be true. This assumption is then used to interpret all experimental results and data. Assigning ages in the millions of years to fossils is less an empirical conclusion than it is a presupposition."[1]

Young-earth creationists point to the discovery of soft tissue remains in dinosaur fossils (and fossil specimens of other creatures) as *prima facie* evidence for the scientific community's philosophical adherence to an old-earth framework, contrary to the evidence. They are quick to highlight how strongly some paleontologists (and other scientists) resist the compelling evidence for soft tissue remnants in fossils—a resistance they claim arises out of a deep concern about the implications of the soft tissue discoveries; namely, that the fossils cannot be millions of years old.

A quick search of the websites of leading YEC organizations, such as Answers in Genesis (AiG), the Institute for Creation Research (ICR), Creation Ministries International (CMI), etc., uncovers a host of popular-level articles describing recent finds of soft-tissue associated with fossils and their implications for the age of the earth. Biologist Brian Thomas, a science writer for ICR, has written—in my opinion—one of the most comprehensive, scholarly reviews of soft tissue finds and, in light of these discoveries, presents one of the best-articulated arguments for the YEC paradigm.[2]

As part of his review, Thomas provides painstaking documentation of a number of key studies that measured the decay rates of collagen under a range of conditions. Based on these research efforts, Thomas concludes that there is absolutely no way that biomolecules such as collagen could survive much longer than hundreds of thousands, let alone millions, of years. With these collagen decay-rate studies at the forefront, Thomas concludes that the fossils can't be millions of years old—if the soft tissue remnants are endogenous. And by all accounts, they seem to be, despite the protests of some paleontologists.

For YECs, the implications are profound. Anderson asserts, "if the soft tissue is original, then a very plausible interpretation is that these dinosaur fossils are actually young (i.e., no more than a few thousand years of age). This younger age contradicts the standard dating methodology used to originally establish the fossils' ages as more than 65 million years. Such contradiction draws the entire geological dating paradigm into question and further exposes the error of accepting evolution as a presupposition for the dating paradigm."[3]

In essence, the argument that Thomas and many other YECs advance depends on the validity of two scientific conclusions: (1) the soft tissue remnants in fossils are authentic, derived from original tissues; and (2) the decay rates of biomolecules—even under optimal burial conditions—are too rapid for these materials to survive for millions of years.

As demonstrated in the last chapter, it does indeed appear that the soft tissue remnants isolated from a wide range of fossilized specimens are authentic endogenous materials, despite the protests of some skeptical paleontologists. The widespread occurrence of a variety of different types of soft tissues, cells, and biomolecules makes it unlikely that these finds are due to contamination—though it is possible that in a few studies contamination may have been a problem. The alternative explanation for the soft tissue finds advanced by paleontologist Thomas Kaye and collaborators does not have much merit, either.[4] Their claim that the soft tissues are bacterial biofilms lacks adequate explanatory power. Their model does not account for the recovery of "blood vessels" from dinosaur remains that were preserved under a variety of different environmental conditions, nor does it explain why the "bacterial films" are hollow, or why they contain microstructures that look like osteocytes outfitted with filopodia.

But what about biomolecule decay rates? Does the breakdown of blood vessels, cells, and organic molecules take place too rapidly for these materials to survive for millions of years? To establish this point, Brian Thomas turned to the work of a diverse cadre of biochemists who have studied the breakdown of collagen found in the mineral matrix of bone. Collagen is the most abundant protein in animals and it is also a highly durable material. Because of these properties, Thomas rightly focuses on collagen breakdown. It represents the most likely biomaterial to survive in fossils.

Some biochemists have characterized collagen breakdown in teeth due to its relationship to dental caries (tooth decay), but most study the decay of bone collagen because this material is used to date archeological sites. Researchers routinely extract collagen from animal remains found near archeological sites and apply carbon-14 dating methods to the bone collagen as a way to determine the age of the finds (assuming that artifacts are less than 50,000 years in age—the upper limit of carbon-14 dating). Knowing how fast collagen degrades in bone is valuable information for researchers who employ this technique.

More recently, biochemists have explored the use of collagen in bone as a source of genetic information about the organisms found at archeological and fossil sites. It is true that DNA is the molecule that harbors genetic information. And researchers have been able to isolate ancient DNA from Neanderthals and other organisms such as cave bears and the woolly mammoth. But DNA is a fragile molecule, which limits its survivability in fossil remains. Collagen is a much more durable molecule, and it is found at extremely high levels in bones. (Twenty-five percent of the dry weight of bone is collagen.) Because the amino acid sequences of proteins (such as collagen) are dictated by the information stored in DNA, the amino acid sequence information of proteins provides *indirect* genetic information. Researchers hope to use collagen amino acid sequences to learn about the natural history of past organisms and their relationship to extant creatures. This interest spurs biochemists' efforts to determine how long collagen can remain intact within a bone matrix. The discovery of soft tissue remains in dinosaur fossils adds to the relevance of this work. And many of these researchers have jumped into the fray over the authenticity of dinosaur soft tissues and the possible mechanisms that promote their survivability for millions of years.

---

Believe it or not: Bone is metabolically active living tissue. This connective tissue is a composite material, consisting of living cells embedded in a matrix made of organic and inorganic components. The organic component of bone (which constitutes about 30 percent of its dry mass) is formed from type I collagen. The inorganic portion is made of salts of calcium phosphate, the most prominent being hydroxyapatite. Bone forms when cells called osteoblasts secrete collagen into the extracellular matrix. The osteoblasts also secrete an enzyme called alkaline phosphatase. This enzyme induces the calcium phosphate minerals to deposit around the collagen fibers. When osteoblasts become trapped inside the matrix they produced (in little openings called lacunae), they become inactive and are referred to as osteocytes. These cells have projections called filopodia that extend away from the cell body. Through the filopodia, the cells interact with other osteocytes and osteoblasts. Presumably the cells communicate with each other through these cellular processes.

Collagen is a fibrous protein material. This biomolecule's basic structural unit is called a triple helix, consisting of three long, extended protein chains that intertwine around each other. At certain points along the triple helix, the individual protein strands are chemically bound to each other to form cross-links. Numerous collagen triple helices assemble in a staggered fashion to form a larger structure called a collagen fibril. Large numbers of collagen fibrils in turn assemble, with the aid of other proteins, into collagen fibers. The collagen fibers give bone its tensile strength. The hydroxyapatite give bone compressional strength. (See figure 4.2, page 56.)

---

When an animal dies, its soft tissues rapidly decay, leaving behind the hard parts of its anatomy—the bones. But over time, the bones also decay. Biochemists think that they have a good understanding of the physical, chemical, and biological mechanisms responsible for the breakdown of bone.[5] Given that bone is a composite material, both the organic and inorganic materials must deteriorate before the bone completely disintegrates.

The primary way that collagen breaks down is through the cleavage of the bonds between the amino acids that make up its backbone. This process generates small collagen fragments that can be carried away from the bone by water that infiltrates the remains. Water drives the cleavage of the collagen backbone via a chemical process dubbed hydrolysis. Temperature and pH influence this reaction. Specifically, collagen hydrolysis occurs more rapidly at higher temperatures and at pH extremes (highly acidic and highly alkaline conditions).

The mineral matrix of bone protects the collagen from breakdown via a variety of mechanisms. For example, collagen molecules form cross-links with the phosphate in the hydroxyapatite mineral phase. These cross-links stabilize the collagen structure. Hydroxyapatite also serves as a pH buffer, preventing extremes in pH. This buffering effect slows down the rate of collagen hydrolysis.

Over time, hydroxyapatite and the other salts that make up the inorganic phase of bones dissolve when exposed to water from the environment. This dissolution process is the primary way that the mineral component of bone disintegrates. As the mineral phase dissolves away, the protection it affords the collagen is lost. This loss leads to an accelerated hydrolysis of the collagen backbone.

The mineral phase also protects the collagen from biodegradation that results from microbial activity and the hydrolytic enzymes that are found in the environment. Once the mineral phase is lost, microorganisms and environmental enzymes will rapidly destroy the organic component of bones. In fact, this is perhaps the fastest acting mechanism responsible for bone deterioration. The process occurs most readily at a neutral pH and moderate temperatures.

---

In his effort to show that collagen decay rates in a bone matrix are much too rapid for this biomolecule to survive for millions of years, Thomas relies on work by researchers from the Interdisciplinary Centre for Ancient Life in Manchester, UK.[6] In one study, the investigators directly addressed the question: how long could collagen survive in fossils?[7] To answer this question, they measured collagen loss in cattle and human bones at 90°C (194°F). The researchers monitored collagen loss at temperatures close to the boiling point of water to gather kinetic data in a reasonable period of time. It still took them about one month to generate the necessary data, even at such high temperatures. They then used this data to calculate the bone loss at 10°C (50°F), which corresponds to the average temperature of an archeological site in Great Britain. The team determined that at these cooler temperatures, it would take somewhere between 200,000 and

700,000 years for 99 percent of the original collagen to decay. This result comports reasonably well with their ability to detect and sequence collagen fragments from the bones of animal remains that date to around 1.5 million years in age.

If this analysis has merit then it sure does not bode well for the claims that collagen fragments could be recovered in dinosaur remains that measure to be 68 million years old. In fact, using the kinetic data they generated for collagen decomposition in cattle and human bone, these researchers questioned whether Schweitzer and her colleagues actually recovered collagen from *T. rex* leftovers. According to the Manchester team, the burial temperatures (20°C) for the *T. rex* specimen unearthed at Hell Creek Formation would have insured the loss of all the collagen (save for 1 percent) on the order of 14,000 years.[8]

According to Anderson, even if the destructive chemical and biological pathways were rendered completely inoperable, there is still no way that biomolecules could survive for millions of years. Environmental radiation would also cause destruction of soft tissue materials in millions-of-years-old fossils.[9] According to Jeffrey Bada at Scripps Research Institute in San Diego, a world authority in amino acid analysis from environmental samples, "Bones absorb uranium and thorium like crazy. You've got an internal dose that will wipe out biomolecules."[10]

Brian Thomas notes that based on thermodynamic considerations alone, no one would expect soft tissue to survive for millions of years. According to Thomas, "If original tissues can avoid being processed by scavengers, microbes, or chemicals, they nevertheless fall apart according to universal entropy, which describes how systems that are left to themselves spontaneously disorganize over time."[11]

---

In an attempt to directly demonstrate that the fossils harboring soft tissue fragments are only thousands, not millions, of years old, Brian Thomas and Vance Nelson, the director of Creation Truth Ministries, sent 14 dinosaur fossil specimens to a commercial laboratory for carbon-14 dating.[12] They reasoned that "Carbonaceous materials from any portion of the geological column deposited millions of years ago should, with the exception of rare instances of contamination, contain zero $^{14}C$ atoms."[13] The 14 samples came from seven dinosaurs, including two from the *Triceratops* horn discovered by Mark Armitage that yielded soft tissue remnants. (See figure 3, page 51.) Radiocarbon was detected in all 14 samples. It should be noted that Thomas and Nelson did not have evidence one way or the other for soft tissue remains in six of the seven samples, with the one exception being Armitage's *Triceratops* horn. The radiocarbon dates for the dinosaur samples ranged from about 20,000 to 41,000 years in age. A bulk sample of the *Triceratops* horn dated to be about 33,500 years in age and the bioapatite fraction dated to be about 41,000 years old.

---

Is the recovery of soft tissue remnants in fossils evidence that the earth is just a few thousand years old?

Anderson notes that "discovery of this preserved tissue has three possible general interpretations: (1) the material is original tissue that was preserved for millions of years by some physical process; (2) the material is original tissue that required no special preservation because the fossil is only a few thousand years of age; or (3) the material is not original tissue but is the product of a microbial biofilm or other contaminating processes that mimic biomolecule-containing tissue."[14]

YECs hold that the first and third scenarios have little scientific support. The soft tissue remnants in fossil fragments appear to be endogenous and bioauthentic. Studies of bone collagen decay indicate that the survival time of soft tissues is only about 1 to 2 million years at best, nowhere near the 60 to 70 million years required for the dinosaur specimens. All that's left is the second option: the fossils are only a few thousand years old. Detection of radiocarbon in several dinosaur fossils seems to add further support for the YEC argument. If the fossils really are just a few thousand years old, then it calls into question the reliability of the radiometric dating methods used to determine the ages of fossils and, consequently, the age of the earth.

Can we trust the reliability of radiometric dating methods? Even if the methods *are* trustworthy, is it reasonable that preservation mechanisms exist that make it possible for soft tissue remnants to survive for millions of years? Or are YECs correct in their assertion that the public has been indoctrinated into believing an evolutionary-based dating of the earth? I take up these questions in the next two chapters.

CHAPTER 3

# Radiometric Dating and the Age of the Earth

One of the first displays visitors encounter when they walk through the doors of *Ripley's Believe It or Not!* Odditorium in Hollywood is the towering mannequin of Robert Wadlow—the tallest man in the world! Wadlow was 8 feet 11 inches tall when he died in 1940 at the age of 22. Wadlow had an arm span of 9 and ½ feet and he weighed 440 pounds. Also on display is one of Wadlow's shoes. He wore a size 37AA shoe that weighed 4 and ½ pounds. As you might imagine, Wadlow didn't buy his footwear at Payless ShoeSource. He had to have them custom made.

Next to Wadlow's mannequin is a giant ruler with the question, "How do you measure up to the world's tallest man?" This question piqued my curiosity during my visit to the Odditorium, and I enthusiastically took my place next the ruler. At 5 feet 11 inches, I am an astounding three feet shorter than Wadlow!

If I told you I knew someone who was over 8 feet tall, you might not believe me because, based on our day-to-day experience, the tallest person we would ever hope to meet would probably be just over 7 feet in height. But, if you saw someone standing next to a ruler and they measured to be 8 feet tall, you would most likely accept the fact that the person was 8 feet tall—even if it defied everyday experience. Why? Because you trust the technique used to measure height. The same is true for weight. You might be hesitant to believe that someone weighed 440 pounds. But, if you saw that person standing on a scale and it recorded their weight at 440 pounds, then you would probably have no trouble trusting that they weighed that much.

But what if I told you someone's age, and you didn't believe me? What could I do to prove it to you? There is no way to measure someone's age in the same way that we can measure height and weight. We *can* measure the passage of time by using clocks. And we can take advantage of this capability to determine someone's age—if we were there to witness that person's birth. But what if we weren't? In that case, we could make use of that person's birth certificate—which certifies the date, time, and location of a person's birth. But what if we have a reason to believe that the document was forged?

---

are they really scientifically determined

YECs are convinced that Scripture teaches the earth is 6,000–10,000 years old. On this basis alone they feel justified in rejecting the scientifically determined ages for Earth and the fossil record. Like most evangelical and conservative Christians (this author included), they regard the Old and New Testaments as the inerrant word of God. However, they consider *their interpretation* of the Genesis 1 creation account to be the only correct reading of the text. Because of their confidence in this interpretation of Scripture, they are unwilling to accept scientific measurements that demonstrate the antiquity of Earth and of the fossil record.

But it's not just a commitment to biblical inerrancy that makes YECs reticent to accept the mainstream scientific view for the age of the earth and the antiquity of its life. Their resistance also has to do with scientists' inherent inability to directly measure the ages of Earth's geological features. Researchers don't have an "age-o-meter" that they can use to directly determine a fossil's age. Scientists can directly measure the size and weight of a fossil. But to determine a fossil's age requires *indirect* methodologies that hopefully employ a reliable "clock" and some way of knowing when the clock got started. YECs are simply not convinced that the scientific community has satisfied either requirement.

The primary way the scientific community dates the earth's geological features and the fossil record is by employing a technique known as radiometric dating. Nearly everyone in the scientific community regards this technique as reliable—but not YECs. These Christians have written countless articles and books detailing reasons why they think radiometric dating is unreliable. Among this list of reasons is the recovery of soft tissue remnants from fossils.

It is true that few, if any, paleontologists ever thought soft tissues would be recovered from fossils. And this discovery has forced them to consider abandoning a long-held paradigm in paleontology about the fossilization process. Yet it has not diminished the scientific community's confidence in the ages assigned to the fossil record, let alone the dependability of radiometric dating. Instead, it has forced them to acknowledge that models for fossilization may be flawed and need revision.

Mary Schweitzer, an evangelical Christian and a former YEC, explains why virtually no one in the scientific community questions the antiquity of Earth or life on Earth. "The fields of geology, nuclear physics, astronomy, paleontology, genetics, and evolutionary biology all speak to an ancient Earth."[1] In other words, there are many independent reasons to think that radiometric dating reliably determines fossils' ages. She is dismayed that YECs use her results to advance the YE paradigm, admitting, "One thing that does bother me, though, is that young earth creationists take my research and use it for their own message."[2]

One goal of this chapter is to offer a few reasons why the scientific community has not wavered in the ages they have assigned to fossils, in spite of the fact that soft tissue leftovers are linked with some of these remains. It is beyond the scope of this book to discuss *all* the scientific evidence for an ancient universe and Earth. Instead, I will refer the reader to a few excellent books that provide scientific and biblical reasons to embrace the antiquity of the universe, Earth, and the fossil record: *A Matter of Days*[3] by astronomer Hugh Ross; *A New Look at an Old Earth*[4] by physicist Don Stoner; and *A Biblical Case for an Old Earth*[5] by physicist David Snoke. Another excellent resources is *The Grand Canyon, Monument to an Ancient Earth: Can Noah's Flood Explain the Grand Canyon?*[6]

The focus of this chapter will be on radiometric dating methods, because this is primarily how scientists determine the absolute ages of fossils. As described in this chapter, geologists obtain radiometric ages on igneous rocks like volcanic ash near the fossils, rather than on the fossils themselves. In this book I will discuss the basics of radiometric dating and then provide reasons why—as a scientist and a Christian—I am convinced that the results of radiometric dating are trustworthy. Readers who want more depth and breadth might consider reading G. Brent Dalrymple's scholarly treatise, *The Age of the Earth.*[7] I refer curious readers who are less adventurous to take a look at Dalrymple's much more accessible work, *Ancient Earth, Ancient Skies.*[8] Or get a copy of his paper, "Radiometric Dating Does Work.[9]" To my knowledge, Dalrymple is not a person of faith, but I have no reason to think he harbors any animosity toward Christianity. To put it another way, Dalrymple has no ulterior motive when he makes the scientific case for the antiquity of the earth based on radiometric dating techniques. The classic work by Davis Young and Ralph Stearley, *The Bible, Rocks and Time,* presents the geological evidence for Earth's antiquity—which includes a discussion of radiometric dating.[10] Both Young and Stearley are evangelical Christians. Finally, I would be remiss if I didn't mention the classic article written by geologist Roger Wiens (also an evangelical Christian), "Radiometric Dating: A Christian Perspective."[11]

As noted earlier, if someone wants to determine my age, all they would need is my birth certificate and a calendar—information about when I was born and a clock that measures the passage of time. A calendar is a clock that is based on the earth's rotation, the Moon's revolution around the earth, and the earth's revolution around the Sun—natural phenomena that all occur in a consistent, periodic manner (at least to first approximation). Radiometric dating is also based on a natural process—the decay of radioactive isotopes—that can serve as a clock.

Isotopes are atoms that have the same number of protons in their nucleus, but a differing number of neutrons. For example, potassium-39 and potassium-40 are both isotopes of potassium. The number of protons in an atom's nucleus (called the atomic number) determines the atom's identity. Any atom with 19 protons is a potassium atom. Both potassium-39 and potassium-40 have 19 protons in their respective nuclei. The difference between the two isotopes relates to the number of neutrons. Potassium-39 has 20 neutrons in its nucleus, whereas potassium-40 has 21 neutrons. The sum of protons and neutrons determines the atom's mass, making potassium-40 slightly heavier than potassium-39. But the number of protons determines an atom's chemical properties, which are identical for the isotopes of an atom such as potassium-39 and potassium-40.

Certain proton and neutron number combinations are unstable. When this instability occurs, the nucleus breaks down through the process of radioactive decay to a stable combination. In this decay process, the (parent isotope) atom's nucleus either gains or loses protons, thereby forming a new (daughter isotope) atom. For example, potassium-40's nucleus is unstable. As a result, the potassium-40 nucleus sometimes captures an electron from the surrounding electron cloud. The electron combines with a proton to form a neutron. The resulting nucleus gains a neutron and loses a proton. Since the total number of protons plus neutrons defines the atom's mass, the atomic mass remains unchanged, but the atomic number decreases by one. The nucleus of the newly formed daughter atom possesses 18 protons and 22 neutrons. Any atom with 18 protons is an argon atom. This transformation, or "radioactive decay"" process, alters the chemical properties of the parent potassium atom, thereby producing a daughter atom of argon, a gas.

There are a number of radioactive isotopes found in nature. The rate that these different isotopes decay, however, varies significantly, with some isotopes decaying more slowly than others. For example, the decay of potassium-40 to argon-40 has a half-life of 1.26 billion years. On the other hand, carbon-14 decays to nitrogen-14 with a half-life of about 5,700 years. Half-life refers to the time it takes for half of the original sample of the parent isotope to decay to the daughter isotope. For instance, if the original sample consists of 1,000 atoms of carbon-14, 5,700 years later only 500 atoms will remain. 11,400 years later, 250 atoms will remain, and so on.

The half-life determined for each individual isotope is a constant because radioactive decay is unaffected by chemical and physical processes and high temperatures and pressures. (It should be noted that *rare* exceptions do exist where chemical and physical processes do alter the radioactive decay rate. These exceptions have no impact, whatsoever, on the validity of radiometric dating.) Radioactive decay involves the nucleus, which is shielded from chemical interactions by the atom's electron cloud. Because of this shielding, chemical processes (which involve the outer portions of the electron cloud) rarely have any effect on radiometric decay rates. Also, the energies involved in radioactive decay are a million times greater than those of chemical activities. The energies are 10,000 to 100,000 times greater than the energies that bind electrons to nuclei. In fact, these energies are so high that geological processes, characterized by intense heat and pressure, are insufficient to influence the radioactive decay process.[12]

Because radioactive decay proceeds at a constant rate, it can serve as a reliable clock to measure the ages of igneous rocks that contain the parent and daughter isotopes. There are over 40 different parent-daughter isotope pairs that can serve as clocks, each one with a different half-life. Each pair represents a distinct radiometric dating method. The parent-daughter pair identifies the radiometric method. For example, the potassium-argon radiometric method uses the potassium-40 parent and the argon-40 daughter isotopes. According to geologist G. Brent Dalrymple:

> Each radiometric method has unique characteristics that make it applicable to particular rocks, particular minerals, or particular geological circumstances. This means that not all the methods can be used on all rocks or under all conditions. It also means that the methods are complementary rather than redundant, with each method working best under somewhat different geological circumstances.[13]

For parent-daughter pairs to be useful as clocks, they must occur in the rocks in measurable quantities and must have appropriate half-lives with respect to the sample. The rule of thumb is that the sample to be dated must fall within an age range that is 0.1 to 10 times the half-life. For example, carbon-14 dating works best if the sample is between 570 to 57,000 years old, because the half-life of carbon-14 is about 5,700 years old.

Thus, radioactive decay provides a reliable clock to age-date igneous rock formations. But how do scientists know when these clocks start? This question is best answered by considering how specific radiometric methods work, such as potassium-argon and carbon-14 dating.

Potassium is one of the most abundant elements in the earth's crust and is found in many different minerals. Three isotopes of potassium exist: potassium-39, potassium-40, and potassium-41. But, potassium-40 is the only isotope of this element that is radioactive, decaying into both argon-40 and calcium-40 by two distinct pathways. When potassium-40 decays, it characteristically produces 11 percent argon-40 and 89 percent calcium-40. The half-life for this process is about 1.26 billion years, making this technique ideal to date samples that are older than a few hundred thousand years in age, though researchers have dated younger samples with the potassium-argon method (and a derivative of this method, called argon-argon dating).

Calcium is a commonly occurring element in many minerals but is not useful for dating rock samples. Geochemists cannot readily distinguish calcium-40 that was already present in the rock from calcium-40 produced by the decay process. On the other hand, argon is not a component of most minerals, making it ideal to age-date rocks. Argon is a noble gas and, consequently, won't bind to other elements. When argon-40 is produced by the radioactive decay of potassium-40, it can get physically trapped in the mineral matrix. But if that matrix melts, argon—a gas—readily escapes the molten rock. Geochemists take advantage of the inert, gaseous properties of argon to age-date rock samples, focusing on quickly cooled lavas.

Because lava is molten, any argon associated with it will readily escape the liquid rock. But once the rock cools, argon produced by potassium-40 decay will be trapped in the matrix of the solid rock. In other words, once the rock cools, the clock starts. Researchers assume that any argon found in volcanic rock was produced after the rock cooled. Measuring the amount of potassium-40 and argon-40 in the rock (taking into account that 11 percent of potassium-40 yields argon-40 and 89 percent produces calcium-40) and applying the half-life of the radioactive decay process makes it possible to date volcanic rock.

Potassium-argon dating is frequently used to date volcanic tuffs—solid rock formed when a layer of volcanic ash is compacted into rock. This material is often found in the geological column and can be used to estimate the age of fossils found above and below the tuffs.

Cosmic radiation impinging upon Earth's atmosphere continually produces the radioactive form of carbon by converting nitrogen-14 into carbon-14. As a result, carbon dioxide in Earth's atmosphere consists of a small amount of carbon-14, with the ratio of radioactive carbon to the stable isotopes of carbon (carbon-12 and carbon-13) being roughly constant. The carbon-14 in carbon dioxide makes its way into living organisms, primarily through photosynthesis. During this process, plants convert water

### How Old Was Volcanic Rock from Mount St. Helens?

YECs will often point to the results of potassium-argon dating of lava rock formed in 1986 as a result of the Mount St. Helens eruption as a reason to question the reliability of radiometric dating.[14] Samples submitted to a commercial lab gave an age of about 350,000 years for whole rock and ages of 340,000 and 2.8 million years for feldspar-glass and pyroxene minerals purified from the rock. This work was published 10 years after Mount St. Helens erupted and should have returned an age of less than 10 years, *if* the potassium-argon dating method was reliable—so claim YECs.

Yet, this result is *not* a sound reason to question the validity of this radiometric dating method. Given that the half-life for the radioactive decay of potassium-40 to argon-40 is 1.26 billion years, this method cannot be used to age-date volcanic rock that is less than a decade old.

But why did these measurements detect any argon-40 at all? Geochemist Roger Wiens explains, "The false radiometric ages of several million years are due to parentless argon . . . first reported in the literature some fifty years ago."[15] This type of argon arises from rocks that are deep within the earth's crust and has a much higher level of argon-40 than atmospheric argon.

and carbon dioxide into carbohydrates. In this way, radioactive carbon becomes incorporated into plants. When animals consume plants, they, too, wind up with carbon-14 becoming incorporated into their tissues. Plants and animals possess a characteristic amount of carbon-14. When an organism dies, it stops taking up carbon-14. Because of radioactive decay, carbon-14 begins to disappear from the organism's remains. The time of the organism's death is the start of the carbon-14 radiometric clock. Researchers know the approximate initial amount of carbon-14 in organisms because of calibration curves prepared from tree rings and sedimentary varves. The half-life of carbon-14 is about 5,700 years, meaning that after about 50,000 years most of the measurable carbon-14 disappears from the organism's remains. Carbon-14 dating is used to determine the age of animal and plant remains (bone and wood), but it also can be used to age-date materials made from once-living organisms such as cloth and paper.

---

So, how do experts know how old a fossil is if we can't put it in an "age-o-meter"? Radiometric dating employs an ensemble of different clocks. And the scientists who perform these methods have ways to determine when the clocks start. The principles of radiometric dating are elegant and straightforward. But, in practice implementing the strategies can be rather involved. Effective use of radiometric dating requires experts who have spent years working with these techniques.

Take potassium-argon dating as an example. While it is generally true that when rock becomes molten the argon gas escapes from the rock, occasionally small amounts of gas can be trapped as tiny bubbles in the rock as it hardens. Argon makes up about 1 percent of Earth's atmosphere. As a result, there will be some argon-40 in the rock that did not result from radioactive decay—a complication that will make it appear as if the sample is older than it actually is. The extraneous argon-40 becomes a much more serious concern for relatively young samples or those with a low level of potassium-40. Fortunately, geochemists know how to correct for the trapped argon-40. As it turns out, there are other isotopes of argon, argon-36 being one. The ratio of argon-40 to argon-36 in the atmosphere is known. So all that researchers have to do is measure the amount of argon-36 in the sample. This value will allow them to calculate how much argon-40 was initially present in the rock when it cooled.

While the principles behind carbon-14 dating are straightforward, application of this method can be tricky as well. Contamination is always a concern. Microorganisms growing on or in the sample will introduce carbon-14 that will skew the ages to be younger than it actually is. Contamination is a real concern for older samples with low levels of endogenous radiocarbon. Another complication has to do with the varying amount of carbon-14 in the atmosphere at different times in recent Earth history and in different locations around the world. Fortunately, geochemists have ways to correct for these differences by comparing the carbon-14 dates of samples of known and varying ages.

The practical challenges that accompany radiometric methods sometimes lead to erroneous results. Hence, YECs cast aspersions on the reliability of these techniques by pointing out instances in which radiometric dating fails. But this criticism is unwarranted. As Dalrymple points out, "A few verified examples of incorrect radiometric ages are simply insufficient to prove that radiometric dating is invalid. All they indicate is that the methods are not infallible. Those of us who have developed and used dating techniques to solve scientific problems are well aware that the systems are not perfect; we ourselves have provided numerous examples of instances in which the techniques fail. We often test them under controlled conditions to learn when and why they fail so we will not use them incorrectly."[16]

---

The principles that undergird radiometric dating are sound. And even though the implementation of those principles can be complicated at times, geochemists possess a good understanding of the source of the complications and how to take them into account and correct for them. But in science, it's not merely enough for a method to be based on sound principles. Practitioners must experimentally demonstrate that it works. That is, the methods must be validated. And this has been done for radiometric methods. For example, in 1997 a team of geochemists from UC Berkeley measured the age of volcanic rock produced when Mount Vesuvius erupted.[17] Pliny the Elder recorded the date that this event took place: August 24, 97 AD—1,918 years before the researchers dated the rock. The research team used argon-argon dating (a variant of potassium-argon dating) on 12 samples taken from the eruption site. They determined that the eruption took place 1,925±94 years ago, demonstrating the reliability of potassium-argon and argon-argon dating.

Scientists have also validated carbon-14 dating. For example, researchers have measured the carbon-14 dates for samples taken from the center of tree rings of bristlecone pines found near the California-Nevada border. These trees can live up to 6,000 years in age. By looking for overlapping tree rings in living and dead trees, scientists can generate a continuous record of tree rings all the way back to 11,800 years ago. Therefore the amount of carbon-14 in a sample of unknown age is compared to carbon-14 in the bristlecone pine tree rings to determine a calendar age in years ago when the animal lived, or charcoal was used from trees.[18]

Another way to validate a method is by cross-correlating its results with results from other techniques. The existence of over 40 different radiometric methods makes this a routine practice among geochemists. It is not unusual for scientists to use several different independent radiometric methods to date rock formations. Dating of meteorites provides a beautiful illustration of how well the results of radiometric methods correlate with each other. Some meteorites are considered to be the oldest objects in our solar system. Meteorites have a complex mineralogy, which affords researchers the opportunity to date them using different radiometric techniques. Meteorites also provide important information about the early history of the solar system. Thus, research groups from around the world study meteorites, which means that most meteorites have been independently dated, time and time again. Virtually every primitive meteorite dates between 4.4 and 4.6 billion years.[19]

Another remarkable illustration of the coherent, consistent results that come from radiometric dating relates to samples used to establish the age of the earth and solar system. As noted, meteorites believed to be the most primitive materials in our solar system date to between 4.4 and 4.6 billion years in age. These dates were generated with a number of different radiometric methods. The same is true for other ancient samples. Scientists have studied meteorites from Mars that date to be around 3.8 billion years in age. Lunar rocks collected during the Apollo missions date to around 4.3 billion years. The oldest rock formations on planet Earth date to around 3.8 billion years. Zircons—found within some of Earth's oldest rock formations—date to between 4.2 to 4.4 billion years. Given that scientists believe that Earth was molten for the first few hundred million years of its existence, and that the Moon formed shortly after Earth formed, these measurements provide us with a remarkably consistent set of data and a remarkably consistent picture of the age of the earth and solar system.[20]

Much more could be written about the reliability of radiometric dating methods. But, on the basis of this short discussion alone, I contend that it is difficult to argue that these tools give scientists anything other than reliable information about the ages of the earth, geological formations, igneous rocks, and hence, fossils. The principles and practices associated with these methods are sound. And they have been successfully validated and cross-correlated.

---

There are quite a few radioactive isotopes with half-lives long enough to make them useful for radiometric dating. But there are many other radioactive materials that are short-lived. While these short-lived isotopes may not have much value for dating features of the geological record, they still give scientists another reason to believe that Earth and the fossil record are ancient. None of these isotopes remain on Earth! Except for carbon-14, beryllium-10, and chlorine-36 (short-lived isotopes continuously produced by cosmic radiation), any radioisotope with a half-life less than 500 million years does not naturally occur anywhere on planet Earth. But

### Why Aren't All Rocks the Same Age?

If YECs are, indeed, correct and the earth is 6,000–10,000 years old, then why don't the radiometric dates for rock samples from all over the world and in different layers of the geological column yield the same age? Biologist Kenneth Miller points out, "If this planet were recently formed . . . potassium-created minerals on a recently created Earth, just like newly formed rocks from active volcanoes, would contain little or no argon."[21] When geochemists determine the ages of volcanic tuffs, the ages of these materials vary widely, instead of yielding one consistent age. Even if the rate of radioactive decay was accelerated at some point, the ages for volcanic tuffs all over the world should be identical. They are not.

there are indicators that these isotopes once existed on Earth because we can find evidence of their daughter isotopes.

Why would these isotopes be absent from the earth? Because Earth is nearly 5 billion years old. Remember the rule of thumb. After a time period that is ten times the half-life, the parent isotope can no longer be readily detected because, for practical purposes, it has all decayed away. These short-lived radioisotopes are like a watch with a dead battery. It *was* keeping time, but then the battery died and the watch stopped.

---

For radiometric dating to provide reliable ages for geological features and the fossil record, radioactive decay rates must be unvarying. And the assumption that the half-life of a radioactive isotope is a constant is more than reasonable. As Dalrymple points out, "Unless there has been some undiscovered change in the fundamental nature of matter and energy since the universe formed, the presumption of constancy for radioactive decay is, for all practical purposes, eminently reasonable."[22]

But what if there *has been* a change in the "fundamental nature of matter and energy"? Since 1997 scientists from ICR have explored the possibility that at different periods in Earth's history (during the first two or three days of the creation week, the curse, and the Noahic flood, for example)

### Does the Bible Teach the Constancy of the Laws of Physics?

YECs argue that the laws of physics changed during the first two or three days of the creation week, the curse, and the Noahic Flood. As a consequence, radiometric decay rates sped up, giving the appearance that Earth is billions of years old, when in fact it is only a few thousand years in age. But is there any biblical warrant for such an idea?

I would argue, "no." In Jeremiah 33:25, God compares his immutability with the constancy of the laws that govern the heavens and the earth. From this passage alone, it is reasonable to conclude that the laws of physics—including radiometric decay rates—have been unchanging since the universe's beginning.

accelerated rates of radiometric decay took place. From their perspective, such seminal events may have given the appearance that Earth is billions of years old (assuming a constant decay rate), when in reality it is only thousands of years old (assuming the possibility of accelerated rates of decay).[23] The ICR scientists dubbed this enterprise the RATE (Radioisotopes and the Age of the Earth) project. Larry Vardiman, who headed this research program states, "It appears that much larger quantities of nuclear decay have occurred in most nuclear processes than would be expected for a few thousand years of radioactivity at the currently observed rates."[24] This is a remarkable admission. If the decay rates are constant and they occur with the half-lives measured today by the scientific community, then even leading YEC scientists acknowledge that Earth must be billions of years old. But to get around this inevitable conclusion, they argue that there were episodes in Earth's history during which the rates of radiometric decay were billions of times faster than today's conventional rates.

But is there any scientific evidence that supports this proposition? If the radioactive decay rates were accelerated a billion-fold, a number of fundamental constants of physics would have had to change to accommodate this phenomenon. There is no evidence whatsoever that the constants of physics changed that radically at any point in time over the last 10,000 years. If the constants of physics did change in the past, astronomers would be able to detect these changes today by observing light emitted from objects that systematically vary in distance from Earth. Because of light travel time, it takes electromagnetic radiation (EMR) about 8 minutes to reach Earth once emitted from the Sun. When astronomers observe the Sun, they are monitoring events that took place 8 minutes ago. When astronomers observe Sirius, they are documenting events that happened 9 years ago. Observations of the Orion Nebula give information that is 1,500 years old. Light detected today from the Cat's Eye Nebula, NGC 869/864, and Cassiopeia reflect events that happened 3,000, 7,000, and 10,000 years ago, respectively. None of the light coming from any of these objects shows any evidence for a change in the fundamental constants of physics.

If the fossils that yielded soft tissue remnants are really millions of years old, then why were Brian Thomas and Vance Nelson able to detect carbon-14 in dinosaur samples?[25] Given that the half-life of carbon-14 is about 5,700 years, all of the carbon-14 should have been extinguished in less than 100,000 years.

Finding carbon-14 in dinosaur fossils does not mean these specimens are only a few thousand years old. There are several ways explain the results. The age Thomas and Nelson reported for the samples (between 20,000 and 41,000 years in age) corresponds to relatively low levels of carbon-14. It is possible that some of the carbon-14 signal stems from contamination of the sample by, say, microorganisms picked up from the environment. They treated their samples to remove any potentially contaminating carbon-14, but the possibility still exists that the protocol may not have removed all the contamination, particularly if the microbes infused into the interior of the specimens.

It is also conceivable that some of the carbon-14 detected by the team is due to a ubiquitous carbon-14 background. Cosmic rays continuously produce radiocarbon. Because of this nonstop production, carbon-14 is everywhere and will show up at extremely low levels in every measurement that is made.

It is also possible that some of the carbon-14 in the fossil samples arises from the conversion of nitrogen-14 to carbon-14 in the fossil interior driven by the decay of radioactive elements, such as uranium and thorium, in the environment. If any soft tissue vestiges are present in the fossils, there will be plenty of nitrogen-14 hanging around. Uranium and thorium from the environment would readily infuse into the fossil interior and as these elements decay, the high energy they release will convert nitrogen-14 to carbon-14.

In reality, any of these possible sources of carbon-14 may be in play at the same time. While any one mechanism may be insufficient to account for the carbon-14 in the dinosaur samples, the combination of sources could readily explain why the YEC investigators detected carbon-14 in the fossil specimens.

One final point: the results Thomas and Nelson obtained for the radiometric dating of the *Triceratops* horn make it impossible that the carbon-14 they detected was endogenous to (produced within) the soft tissue.[26] Recall that Armitage and his associate Kevin Anderson (both YECs) uncovered soft, flexible brown sheets about 8 inches by 4 inches in size from the *Triceratops* fossilized horn after soaking pieces of it in a mild acid bath for a month.[27] (This procedure provides a source of carbon-14 because of the small amounts for carbon dioxide dissolved in the liquid.) That is a considerable amount of soft tissue. If the *Triceratops* horn were somewhere between 3,000 and 6,000 years old, then there should have been somewhere between 50 and 75 percent of the original carbon-14 in the sample. This amount of material should have easily produced a strong carbon-14 signal. The fact that they measured the age of the bulk *Triceratops* horn at about 33,500 years in age and the bioapatite fraction around 41,000 years old means that less than 2 percent of the original carbon-14 was present in these samples, if the results of this measurement are taken at face value. This outcome makes no sense if these samples are less than 6,000–10,000 years old. (See figure 3.)

It is also strange that the bioapatite fraction (which should have an enriched organic content) should be dated 10,000 years older than the bulk horn. These spurious results make sense if the carbon-14 signal was due to contamination, ubiquitous background radiocarbon, and/or irradiation of the flexible sheets by uranium or thorium. It is impossible to reconcile the data Thomas and Nelson reported with any scenario that would treat the flexible sheets of soft tissue as only 3,000 to 6,000 years old.

But YECs would likely point out that the discrepancy in the carbon-14 dates reflects the acceleration in radioactive decay rates that took place at the fall of Adam or during Noah's flood. But this rejoinder doesn't solve their problem. In fact, it highlights another inconsistency in their model. For the sake of argument, let's assume that the *Triceratops* horn is 3,000 years old but measures to be 30,000 years old because of accelerated radioactive decay. This differential means that the rate of radioactive decay must have sped up by a factor of ten. But, the RATE scientists contend that radioactive decay rates during select times in Earth's history were *billions* of times faster than today, not merely an order of magnitude faster. If the radioactive decay rate increased by a magnitude factor of billions, then there should be no carbon-14 in any of the fossil samples. The presence of carbon-14 gives scientists another reason to think that the carbon-14 signal is due to contamination, background carbon-14, and/or conversion

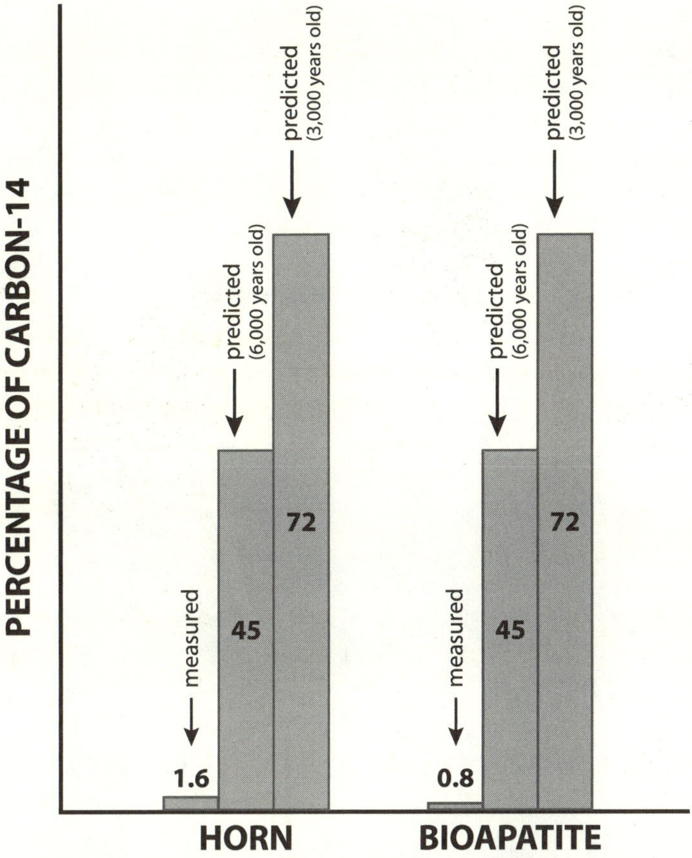


**Figure 3: Carbon-14 in *Triceratops* Horn Samples**

from nitrogen-14 by exposure to radioactive materials such as uranium and thorium.

Based on this cursory treatment of radiometric dating, it should be evident why nearly all scientists think the technique is reliable. Even though soft tissue remnants have been recovered from fossil remains of organisms that date to almost 500 million years in age, there is no reason to doubt the fossils' ages, as determined by radiometric dating. Nevertheless, we are still left with the intriguing scientific question as to how it is possible that soft tissue can survive that long. Can we believe it or not? I take up that query in the next chapter.

CHAPTER 4

# How Did Soft Tissues Survive in Dinosaur Fossils?

The day I bought a Rubik's Cube is the day that will live in infamy—at least for me. This toy was the rage my junior year in high school. Several of my friends had a Rubik's Cube. We would bring them to school and work our rotating three-dimensional puzzles between classes. A few of my buddies were adept at solving the cube. I wasn't. This stupid toy caused me no end of frustration. I still haven't solved a Rubik's Cube to this day.

My inadequacies with a Rubik's Cube undoubtedly influenced my reaction to one of the more remarkable displays in the Hollywood *Ripley's Believe It or Not!* Odditorium: a portrait of the singer and actor, Frank Sinatra—made from 450 Rubik's Cubes! Entitled "Blue Eyes," this mosaic was designed and created by Josh Chalom. Not only did Chalom have to solve 450 Rubik's Cubes but also he had to twist and turn each cube so it formed a "just-right" pattern that fit with the patterns of the other cubes

### Speedcubing

Some people aren't satisfied by solving a Rubik's Cube. They have to do it faster than anyone else. The *Guinness World Records* organized the first speedcubing competition in Munich, Germany, in 1981. Jury Froeschl was the winner, solving the Rubik's Cube in 38 seconds. His time was a mere snail's pace compared to Minh Thai's time of 22.95 seconds. Thai won the first international world championship, held in Budapest, Hungary, in 1982. In 2015, Feliks Zemdegs solved a Rubik's Cube in 7.56 seconds in Sao Paulo, Brazil, at the Rubik's Cube World Championship. And in November 2015, Lucas Etter solved a Rubik's Cube in 4.90 seconds.

in order to produce a remarkably detailed portrait, flush with depth and shadowing.

---

Sometimes solving scientific problems can be a lot like working a puzzle. This is clearly the case when it comes to explaining how and why soft tissue fragments survive in fossils—in some cases up to 500 million years. YECs claim that the survival of original tissue materials makes it unlikely that these fossils are millions of years old. Yet, there is no reason to think that the radiometric dating method used to determine the ages of fossils is inherently faulty. In fact, there is every reason to think these dating techniques are reliable. If so, then how do biologists explain the survival of soft tissues in fossils?

As it turns out, the scientific community has proposed a number of different mechanisms they think could account for the persistence of tissue, cellular, and biomolecular vestiges in fossil remains. In reality, it is unreasonable to think that any *one* mechanism is sufficient on its own to explain the survival of soft tissue remains. More than likely, a *combination* of factors worked in conjunction to preserve the original tissues in the fossils, with each one functioning like a piece of the puzzle—just like the 450 uniquely solved Rubik's Cubes that Chalom used to piece together Sinatra's portrait.

---

Generally speaking, it is true that organic molecules are rather "delicate" materials that readily break down. But not all molecules are the same. Some organic compounds form extremely stable structures that result in hard and durable materials. This is the case for most of the biomolecules that survived in the fossil remains of dinosaurs and other creatures. In many instances, the soft tissue materials that survived in fossils are made up of molecules with one of two properties: (1) an extensive cross-linking; or (2) a chemical makeup similar to graphite, which is the most thermodynamically stable structure possible for an organic molecule.

Cross-links refer to bonds that link one polymer to another. (A polymer is a chain-like molecule made of subunit molecules that are joined together. For example, proteins are polymers made of amino acids.) There are different types of cross-links. Some, such as hydrogen bonds, are rather weak and impermanent. Yet, collectively these weak cross-links can impart structural stability through a phenomenon known as cooperativity. Other cross-links are also impermanent, but much stronger, such as the electrostatic bonds that form between oppositely charged residues in polymer chains. The most stable and long-lasting cross-links are covalent bonds between atoms located in two different polymer chains.

Polymers readily break down if the chemical bonds that form their backbone are cleaved. When the bonds between subunits break, the polymer fragments diffuse away from each other. Over time, this process converts long polymer chains into short molecular fragments. But when polymer chains are cross-linked, they become more enduring because the cross-links prevent the polymer fragments from diffusing away from each other. Depending on the chemical identity of the subunits, sometimes the broken backbone bonds of the polymer can reform, reversing the damage. In essence, the cross-links provide structural reinforcement. The more heavily cross-linked (or reinforced) a polymer is, the more resistant it is to chemical breakdown.

Chemical materials with graphite-like structures are also incredibly stable. Graphite is also a polymeric material, of sorts. The basic structural unit of graphite consists of a ring made up of six carbon atoms. The atoms in the ring take part in a special type of chemical bond called an aromatic bond. This type of bond is characterized by extensive electron delocalization around the ring. The greater the degree of electron delocalization in a molecule, the more stable it is. Electrons are like energetic little kids—they don't like to stay in one spot. They much prefer to run around with abandon. In a typical chemical bond, the electrons are localized between two atoms. But in an aromatic bond, the electrons can "run around" the entire ring, just like kids on a playground.

Graphite forms when the six-membered carbon rings fuse together into a massive two-dimensional sheet. (See figure 4.1.) Fusing the rings together makes the "playground" bigger for the electrons. They can run from ring to ring without any restrictions. The larger the graphite sheet,

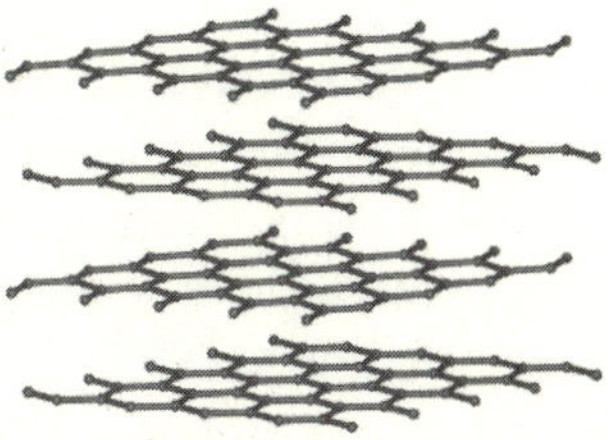

**Figure 4.1: The Structure of Graphite**
Michael Ströck/Wikimedia Commons

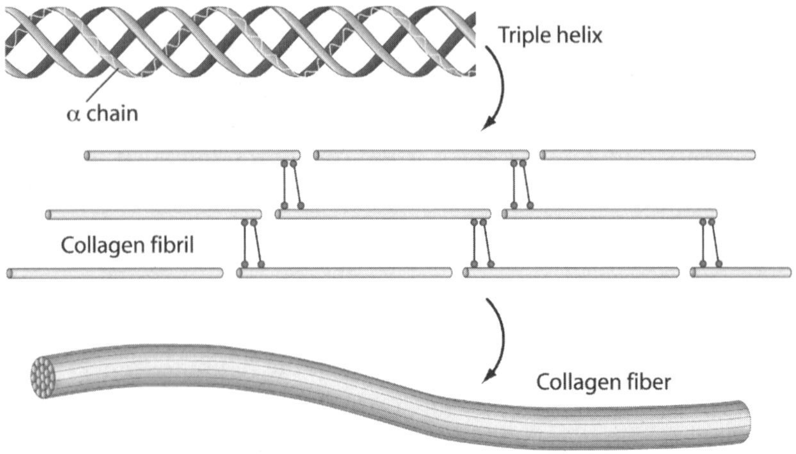


**Figure 4.2: The Structure of Collagen**

the more stable the structure, because larger sheets permit greater electron delocalization. In fact, the electron delocalization is so extensive in graphite that it is one of the few organic materials that conduct electrical current.

The following list describes seven durable chemical structures. With these simple chemical principles in mind, it begins to become apparent why certain soft tissue materials could have survived for millions of years.

*1. Collagen.* This biomolecule is a fibrous protein. Collagen's basic structural unit is called a triple helix, consisting of three long, extended protein chains that intertwine around each other. At certain points along the triple helix, the individual protein strands are chemically bound to each other to form cross-links. Numerous collagen triple helices assemble in a staggered fashion to form a larger structure called a collagen fibril. In turn, large numbers of collagen fibrils assemble, with the aid of other proteins, into collagen fibers. (See figure 4.2.)

The intertwining of the collagen protein chains and fibrils along with the extensive cross-linking between the collagen chains makes collagen an incredibly durable material. This property makes collagen well-suited to form the connective tissues found in animals. For that reason, it isn't surprising that some collagen fragments would survive in fossilized dinosaur bones. Even if the individual protein strands break down, the fiber would still remain largely intact because of all the association points. Once the protein strand breaks, the fragments are held in close proximity by the contact points. This forced closeness allows for broken strands to occasionally rejoin and reform the original protein. If the broken strands were not held juxtaposed to each other, the fragments would diffuse away from each other, thus preventing the reversal of the degradation process.

*2. Keratin.* This protein is known as an intermediate filament. Its basic structural unit is an extended fibrous protein chain. The chain will intertwine with another to form a dimer. Two dimers, then, interact to form tetramers, and the tetramers interact to form octamers. These octameric complexes are referred to as keratin filaments. Keratin filaments intertwine to form coils. As one of their characteristic features, keratins possess a high level of cysteine, the sulfur-containing amino acid. (Hair is largely comprised of keratin. Cysteine makes up 14 percent of human hair, which is why the keratin in hair smells so pungent when it burns.) When incorporated into a protein chain, cysteine can react with another cysteine residue to form a special type of covalent cross-link called a disulfide bridge. These cross-links are the same type of cross-links that make vulcanized rubber heat resistant and rigid. The more disulfide bridges, the more durable the materials made up of keratin. The cross-links also make keratin insoluble in water, affording it protection from hydrolysis.

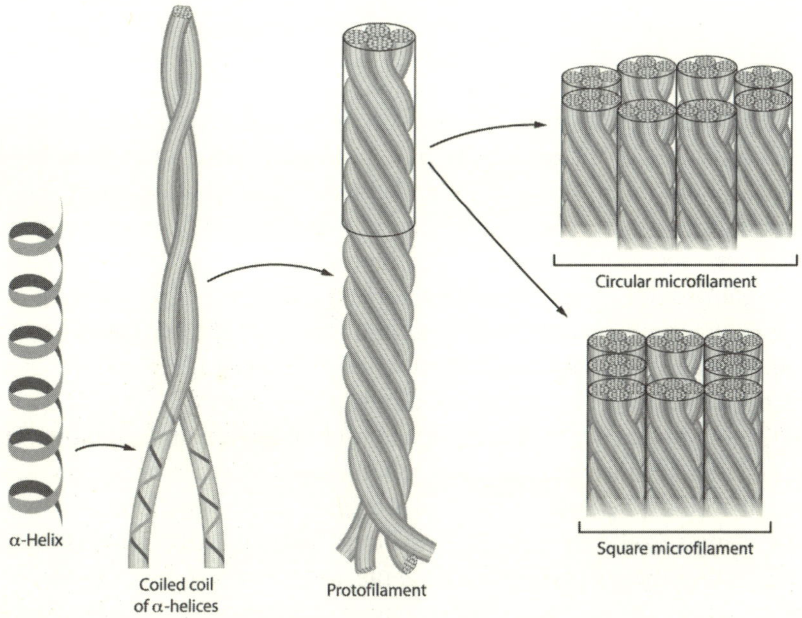


**Figure 4.3: The Structure of Keratin**

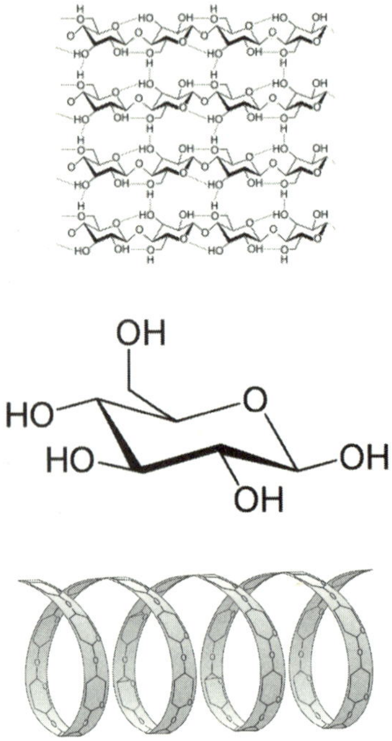

**Figure 4.4: The Structures of Starch and Cellulose**

(From top to bottom) Laghi/Wikimedia Commons; Yikrazuul/Wikimedia Commons; H. Hoffmeister/Wikimedia Commons

Keratin makes up hair, feathers, skin, fingernails, reptile scales, animal claws, birds' beaks, tortoise shells, porcupine quills, and on and on. Given how hard and tough materials from keratin are, it is not unreasonable to think that this material could persist in fossilized feather-like structures linked with dinosaurs or in fossilized reptile skin.

*3. Chitin.* Some organic materials are remarkably robust—chitin is one of them. Chitin is a sugar (more specifically, a carbohydrate). Some of the more familiar sugars, such as starch, break down quickly. However, other sugars, like cellulose, are extremely durable materials.

Chains of glucose molecules make up both starch and cellulose. The key difference between starch and cellulose is the nature of the chemical bond that joins the glucose molecules together. The bond found in starch is called an α-1,4 linkage, while the cellulose bond is called a β-1,4 linkage.

Whether the glucose units are connected via an α-1,4 or a β-1,4 linkage makes all the difference to the sugar's durability. When joined together with α-1,4 bonds, the linear glucose chains adopt an open helical structure. But if they are bound by a β-1,4 linkage, the chains exist in an extended state. In this latter configuration, the chains also interact with one another via hydrogen bonding, resulting in multiple cross-links between the chains. Multiple chains interact in this way to form a two-dimensional network, or "sheet," of cellulose chains (figure 4.4). This large number of cross-links explains cellulose's robustness.

Chitin has a similar structure to cellulose, except that its chains are made up of N-acetylglucosamine subunits joined together with β-1,4 linkages (figure 4.5). The β-1,4 linkages in chitin give it structural properties and durability similar to those of cellulose, with one exception: the N-acetyl chemical group creates more opportunities for hydrogen bonds to form between chains. This means that the cross-linking in chitin is more extensive than in cellulose, thus making it even more durable.

Chitin's heavily cross-linked structure makes it reasonable to conclude that this molecule could survive for a long period of time. It is not outlandish to think that chitin could persist for up to 500 million years under the right conditions. In fact, chitin is such a durable substance that engineers are exploring how it can be used to develop new types of materials that can withstand high temperatures and pressures.[1]

*4. Eumelanin.* Given its chemical structure, it is not surprising that eumelanin can persist for over 160 million years in fossilized ink sacs. Eumelanin is a polymer made up of subunit molecules with a chemical bond between them that leads to extensive electron delocalization along the entire backbone of the polymer. This property makes the eumelanin polymer extremely durable and difficult to break down. Moreover, the individual polymer chains of eumelanin are extensively cross-linked

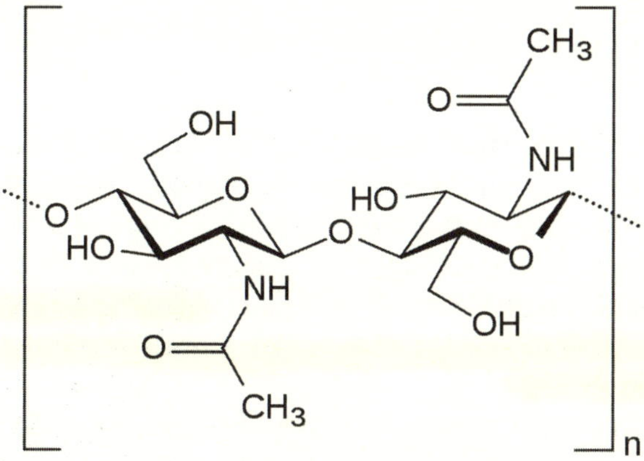


**Figure 4.5: The Structure of Chitin**

Dschanz/Wikimedia Commons

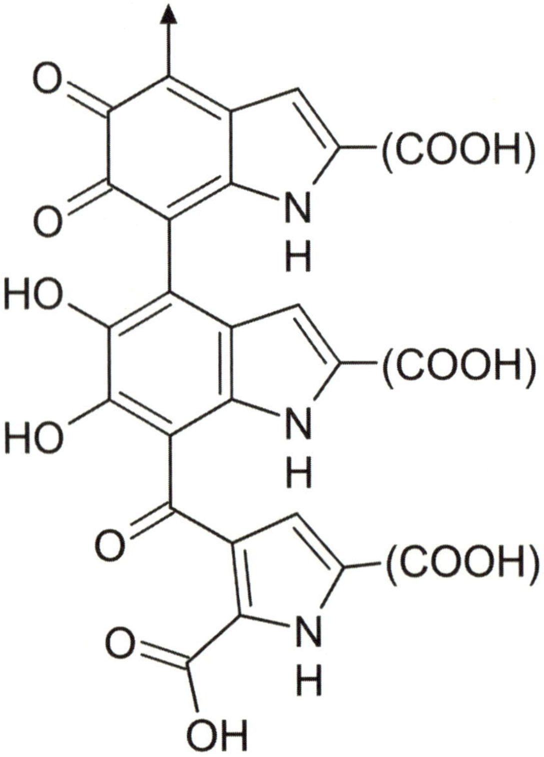


**Figure 4.6: The Structure of Eumelanin**

Roland Mattern/Wikimedia Commons

with each other, making this material even more resistant to chemical decomposition. The researchers who discovered eumelanin in the fossilized ink sac of a 160-million-year-old cephalopod fossil explained that, "Biomolecules that are polymeric and highly cross-linked in their original state exhibit the greatest resistance to alteration. Melanin, a complex biopolymer, meets this high resistance criterion."[2] Considering its chemical structure, it is not unreasonable to think that eumelanin persisted for well over 160 million years.

5. *Heme.* This small molecule binds to hemoglobin, in turn, serving as hemoglobin's oxygen-binding site. Two components make up heme: a large-ringed structure called porphyrin and an atom of iron. The iron resides within the cavity formed by the porphyrin ring. The ring itself is formed from four pyrrole molecules. These molecules consist of a five-membered ring that includes four carbon atoms and a single nitrogen

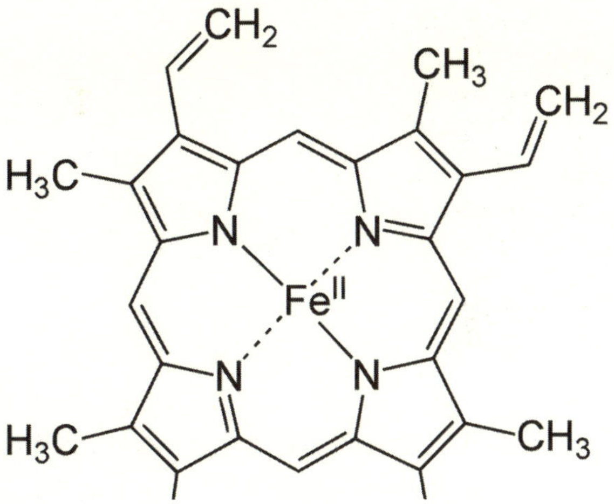


**Figure 4.7: The Structure of Heme**
Yikrazuul/Wikimedia Commons

atom. The five atoms are joined together by aromatic bonding. Because of their aromatic nature, pyrrole molecules are chemically stable materials. The four pyrrole rings are joined to one another by a chemical linkage called a methine bridge. This bridge allows the delocalized electrons of each pyrrole ring to freely move across the bridges from ring to ring, making the entire porphyrin ring aromatic. In other words, the porphyrin ring is an extremely stable compound, which helps explain its presence in fossilized dinosaur bones.

*6. Quinones.* These small compounds consist of either a single ring made up of six carbons, or two or more fused six-carbon rings. One of the rings has two oxygen atoms bound to it through double bonds. The double bonds of quinones are conjugated. This property allows for extensive electron delocalization, just like aromaticity. In other words, quinones are also extremely stable compounds, which helps account for their preservation in 340-million-year-old sea lilies.

*7. Blood vessels.* These extremely durable materials are made of endothelial cells that organize into a channel. An elastin matrix, basement membrane, muscle fibers, and a collagen matrix, respectively, surround this channel. The design makes it possible for blood vessels to endure high pressures that result when blood is pumped throughout the body. Given how durable blood vessels must be, it is reasonable to think that they would be highly resistant to break down. The resilient, durable nature of blood vessels helps account for their survival in dinosaur remains.

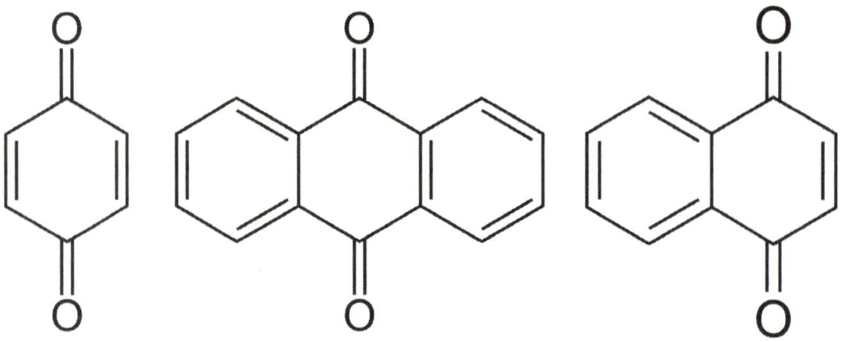

**Figure 4.8: The Structure of Typical Quinones**

While durable chemical structures help explain why some biomolecules and tissues can survive for vast time periods, durability alone is not sufficient to account for the survival of soft tissues in fossil remains for upwards of hundreds of millions of years. Many other conditions must also be met simultaneously. The most important of these relates to the rate of fossil formation. Fossilization must occur more rapidly than the processes that degrade the soft tissue materials, if these materials are to be preserved. Scientists have identified at least nine stabilizing mechanisms that can prevent soft tissues from degrading before fossilization takes place.

*1. During fossilization, mineral-rich water infuses the remains of the organism.* As this infusion takes place, the original minerals in the bone (and other parts of the remains) are replaced with minerals from the environment. Schweitzer and her colleagues contend that when these environmental minerals encounter the stabilized soft tissues, they will precipitate onto the surfaces of the preserved tissues, much in the same way that hydroxyapatite precipitates onto collagen surfaces during bone formation.[3] As a result, the precipitated minerals entomb the soft tissues. This entombment serves to protect the soft tissue remains from water, oxygen, environmental enzymes, and microbes, the most likely causes of destruction.[4]

*2. Burial conditions also appear to be important.* For example, Schweitzer and her team surveyed the prevalence of soft tissues in a number of fossil remains recovered from different geological settings (such as fluvial deposits, sandstones, cave deposits, mudstones, and marine deposits). They discovered that fossils unearthed from fluvial and sandstone environments were much more likely to have soft tissue remnants than specimens retrieved from mudstone and marine settings.[5] Presumably, water more readily drains away from animal remains located in fluvial deposits and

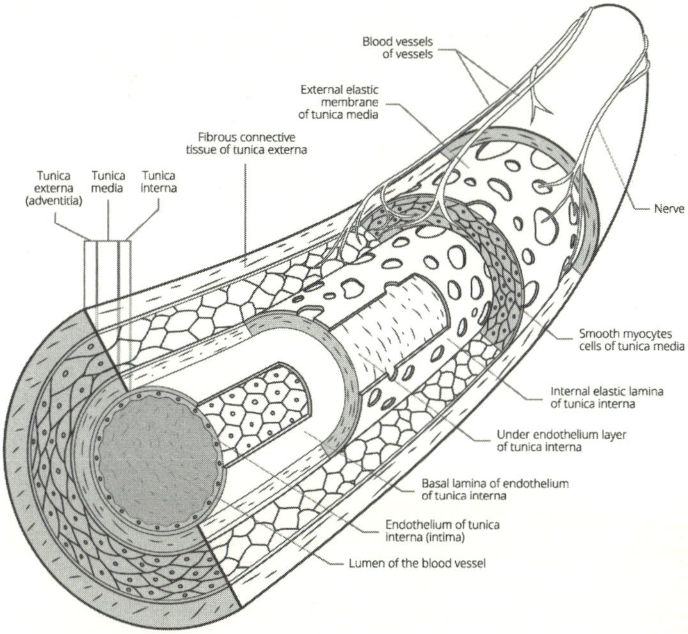


**Figure 4.9: A Cross Section of a Typical Blood Vessel**

iStock

sandstones, creating drier conditions, removing microbes and environmental enzymes.

Researchers believe that burial conditions also help explain the persistence of quinones in 340-million-year-old sea lilies.[6] According to the scientist who discovered the ancient quinones, the fossils experienced a rapid burial that isolated the organic and inorganic materials connected to the sea lily remains from fluids moving through the rocks, preventing their chemical alteration. Also, the rock formation that housed the fossils escaped metamorphosis, which would also have altered the fossils and their associated organics. Additionally, the physiology of the sea lilies promoted stability of the quinones. Sea lilies have a porous skeleton. Some of the pores are filled with organic materials. When the organism died, the pores were sealed with calcite, thus trapping the molecules in a protected environment.

*3. Dry, anhydrous conditions are also necessary to preserve soft tissues.* If water isn't present, then the hydrolysis of biomolecules cannot take place. Ironically, in some instances, a limited amount of water may actually help preserve biomolecules such as collagen. When partially hydrated, the three-dimensional structure of collagen becomes altered (as a result of the hydrophobic effect). This alteration makes it harder for water molecules to reach the collagen backbone and also strengthens the cross-links between collagen chains, preventing the breakdown of this molecule.[7]

Researchers from South Africa argue that extreme dehydration of animal remains (such as would occur during a severe drought) may help to preserve collagen long enough so that mineral entombment can take place.[8] These investigators demonstrated that collagen was structurally intact when isolated from decomposing dolphin, python, and turtle tissues that had undergone dehydration. They also showed that when the tissues were rehydrated, the collagen displayed minimal evidence of damage.

4. *Sequestering soft tissues from oxygen ranks as another significant preservation factor.* Oxygen is a highly reactive, chemically destructive material that readily destroys organics through a chemical process called oxidation. Though counterintuitive, reactions between oxygen and soft tissues may actually help preserve remnants of these materials. When oxygen reacts with organic materials it can generate free radical species. These molecular entities are highly reactive and will lead to the formation of cross-links between the components of cell membranes (such as lipids), making them resistant to break down.[9] When an organism is alive, these kinds of free-radical induced cross-linking reactions are disastrous. But when no longer living, the chemical processes help preserve soft tissues by forming stabilizing bonds. This type of cross-linking helps explain why remnants of more "delicate" materials, such as red blood cells and osteocytes, persist in dinosaur remains.

5. *To be preserved, soft tissues also must be kept away from environmental enzymes and microorganisms.* The environment is chock-full of digestive enzymes that will break down the proteins that make up soft tissues, such as collagen. Likewise, microbes will use soft tissue materials as a food source, rapidly consuming them. But even though enzymes are highly destructive, their activity can be inhibited in a variety of ways. For example, collagenases (enzymes that degrade collagen) have a hard time physically gaining access to collagen when this protein is associated with a mineral matrix such as hydroxyapatite.[10] Without this access, the enzymes can't chew up this protein.

Heme (the oxygen-binding component of hemoglobin) can also inhibit enzymatic activity.[11] At the time of the dinosaur's death, heme would be present in bone at relatively high levels, because red blood cells located in the blood vessels permeating bone are loaded with heme-containing hemoglobin.

The soft tissues also have to be protected from microbial attack in order to persist in fossil remains. As it turns out, hemoglobin possesses the capacity to kill microorganisms.[12] This antimicrobial activity appears to be linked with the protein chains that make up the hemoglobin molecule. The antimicrobial activity of hemoglobin imparts the blood within the bones of animal remains with the capability of keeping microorganisms at bay while the fossilization and entombment process takes place.

Schweitzer and her associates think they have discovered another way such an attack can be thwarted. The iron derived from the heme component of hemoglobin also appears to play a role in preserving the soft tissue from microbial destruction.[13] This element plays many critical roles in living organisms. But in its free form, it can cause all kinds of problems. Animals employ protective mechanisms to keep free iron from causing damage while exploiting its life-enabling properties. One method relies on proteins to bind and isolate the free iron from other tissues. Hemoglobin ties up most of the iron, with other proteins (such as non-heme iron proteins and ferritin) binding the remaining fraction. When an animal dies, however, the iron-sequestering mechanism fails, turning the iron loose. In its free form, iron is highly destructive to living tissues, and would also inhibit microbial growth during the fossilization process.

*6. But ironically (pun intended), iron may also help stabilize soft tissues from chemical destruction, as well.* Iron exerts its destructive effect by catalyzing the production of reactive-oxygen species. In the process, this metal transforms into a mineralized precipitate (rust). As discussed, the reactive-oxygen species cause reactions that cross-link proteins and membrane components. These reactions would stabilize soft tissue remains with the cross-linking agents that serve the same function as tissue fixatives (such as formaldehyde), thereby preserving vessels, cells, and molecules.

Schweitzer and her team noticed that iron particles have been found in close contact with soft tissue fragments in dinosaur fossils. This discovery serves as a telltale sign that iron has played a role in fixing and preserving the soft tissue remains in these fossils. To confirm that hemoglobin in dinosaur bone really does exert a protective effect, Schweitzer's team soaked blood vessels isolated from ostrich bone in two different solutions: one containing hemoglobin and one consisting of pure water. After two years, the blood vessels in the hemoglobin bath remained intact, but the ones in water fell apart. So, it looks as if iron does indeed play a role in preserving soft tissues in dinosaur bones after they die.

*7. Temperature also plays a role in soft tissue survival.* The rates of chemical reactions roughly double each time the temperature increases by 10°C (18°F). In other words, the higher the temperature, the faster chemical degradation takes place. Because of this relationship, soft tissue remnants are more likely to survive for longer periods of time if the fossil is in a cooler environment. Cooler temperatures also retard the activity of microorganisms.

On the other hand, high temperatures can also contribute to soft tissue survivability. Heat can sterilize the fossil remains by killing off microbes and inactivate enzymes thereby denaturing them. Even though heat speeds up the breakdown of molecules such as collagen, it can also accelerate the formation of cross-links between proteins during the degradation process, which arrests the destruction of soft tissue.

*8. Additionally, pH contributes to soft tissue preservation.* Proteins are less likely to break down when the pH of their surroundings is near neutral. But neutral pH provides ideal conditions for enzymatic activity and microbial growth. Highly acidic and highly alkaline conditions promote the hydrolysis of protein backbones. But, pH extremes also deactivate digestive enzymes and keep the growth of microorganisms in check.

*9. Minerals also play a key role in preserving soft tissue materials.* As noted, collagen embedded in the mineral matrix of bone is afforded protection from digestive enzymes such as collagenases, because the enzymes can't physically access the collagen when this molecule is connected to bone. The hydroxyapatite of bone also prevents the chemical breakdown of collagen. Some of the amino acid constituents of collagen form chemical bonds with the phosphate moieties of hydroxyapatite.[14] These bonds serve as cross-links, holding the collagen molecule together, even when some of the chemical bonds that form the collagen backbone become cleaved.

Researchers also think that interactions between the mineral aragonite and chitin help account for the survival of this organic material in the 35-million-year-old cuttlefish remains recovered in Mississippi.[15] (See chapter 1, pages 23–24.) The interactions between calcium carbonate minerals (calcite and aragonite) and shell-binding proteins most likely accounts for the survival of these proteins in the 15-million-year-old fossilized shells of the gastropod *Ecphora*.[16]

The bottom line: the scientific community has proposed a number of distinct mechanisms to explain how soft tissue materials survive in fossils. No single mechanism can fully account for the persistence of soft tissues, but in combination they can. It is also important to recognize that when it comes to soft tissue survival, things aren't black and white. Sometimes pathways that are destructive under certain sets of conditions can be protective in other circumstances.

But YECs are not convinced. For example, Brian Thomas argues that the laws of thermodynamics alone demand that soft tissue materials would degrade. Thomas states, "If original tissues can avoid being processed by scavengers, microbes, or chemicals, they nevertheless fall apart according to universal entropy, which describes how systems that are left to themselves spontaneously disorganize over time."[17] Thomas is, indeed, correct. Left on their own, organic materials will eventually break down into carbon dioxide, nitrogen, and water because of the effects of entropy. Thomas seems to imply that this process should be well under way, if not completed, after 70 to 80 million years. Unfortunately, this is where Thomas is mistaken. His error: failing to distinguish between thermodynamically controlled and kinetically controlled chemical reactions.

Chemical thermodynamics is the area of chemistry that focuses on the energy state of the molecules in a chemical system. In order for a chemical reaction to proceed, the products have to be at a lower energy state (as measured by a quantity known as the Gibbs free energy) than the reactants. If this energy differential is the case, the reaction is said to be spontaneous, meaning that the reaction will take place. And there is nothing anyone can do to stop it. On the other hand, if the products are at a higher energy state than the reactants, then the reaction will not occur. And there is nothing that anyone can do to cause a reaction that is thermodynamically unfavorable to transpire.

While thermodynamics can tell us *if* a reaction will or will not occur, it cannot tell scientists anything about *when* it will happen. That is the purview of chemical kinetics. Some chemical reactions proceed rapidly and others slowly. A quantity known as the energy of activation determines the rate of chemical processes. The energy of activation reflects the energy barrier reactants have to overcome before they can be converted into products: the greater the energy of activation, the slower the reaction; conversely, the smaller the energy of activation, the more rapidly a reaction will proceed.

Due to the relationship between thermodynamics and kinetics, it is possible that a chemical reaction can be highly favorable from a thermodynamics standpoint, but extremely slow because the energy of activation is quite high. In fact, if the energy of activation is extremely high, the reaction may never take place, practically speaking. An example of a kinetically controlled chemical process is the transformation of diamond into graphite. Both materials are different forms of elemental carbon. From a thermodynamics standpoint graphite is at a much lower energy state than diamond and, therefore, all the diamonds in the world should spontaneously transform into graphite. But they do not and they will not. Why? Because the energy of activation is so large that this spontaneous chemical reaction won't happen on any appreciable time scale.

This same principle applies to soft tissue remnants interned within fossils. The cross-linking reactions and interactions with replacement minerals will create altered versions of the original soft tissue that possess a high-enough energy of activation such that the spontaneous decay of soft tissues into carbon dioxide, nitrogen, and water does not occur.

---

But what about the study that determined that, at best, only 1 percent of the original collagen in bone remains after 700,000 years?[18] How is it possible to square this result with the recovery of soft tissue vestiges in fossils that are millions of years old? Recall that researchers from the University of Manchester, in the UK, measured collagen loss in cattle and human bones at 90°C (194°F). (See page 33.) The researchers followed collagen loss at temperatures close to the boiling point of water to accelerate the decay process. It still took them about one month to generate the necessary data, even at these high temperatures. They then used this data to calculate the bone loss at 10°C (50°F), which corresponds to the average temperature of an archeological site in Great Britain. These calculations made use of the Arrhenius equation. This equation allows scientists to calculate the rate for a chemical process (such as the breakdown of collagen) at any temperature, once the rate has been experimentally determined for a single temperature. The only assumption is that the physical and chemical properties of the system are the same as the temperature used to measure the reaction rate and the temperature used to calculate the reaction rate. If the conditions differ, then a phenomenon known as an Arrhenius plot break occurs. This discontinuity makes it impossible to calculate the reaction rate.

One has to wonder if the kinetic data for collagen breakdown in bone near the boiling point of water is relevant to the breakdown rate for temperatures that would be under 100°F, let alone to temperatures near 50°F. It is quite possible that at such high temperatures, the collagen would undergo structural changes (for example, breaking of interchain hydrogen bonds that cross-link collagen chains together) that would make it much more susceptible to chemical degradation than at lower temperatures where collagen would remain in its native state. In other words, the conditions employed by the research team from the University of Manchester may not be relevant to Schweitzer's discoveries (and finds of other paleontologists).

Another reason why the results obtained by the Manchester team do not contradict the recovery of collagen from 70- to 80-million-year-old dinosaur fossils relates to the question that this group addressed. Namely, how long can collagen last in animal remains in a form that can be isolated and used as a source of genetic information about the organisms found at archeological and fossil sites? In other words, the team wasn't merely interested in how long chemically and physically altered collagen fragments would persist in fossil remains. Rather, they focused on how long collagen will retain a useful form that can yield insight into the natural history of past organisms.

Specifically, they were interested in the survival of "the non-helical collagen telopeptides located at the very ends of each chain and recently considered potentially useful for species identification in archeological tissues."[19] One difficulty, the researchers lament, is that this region of the collagen molecules is "lost to the burial environment within a relatively short period of geologic time."[20] As they point out, the parts of the collagen molecule most useful to characterize the natural history of past organisms and their relationships to extant creatures, unfortunately, are "regions of the protein that do not benefit from as many interchain hydrogen bonds as the helical region and thus will likely be the first to degrade."[21]

Thus, when these scientists claim that only 1 percent of the original collagen remains between 200,000 and 700,000 years, they are probing intact material that corresponds to the most labile (unstable) parts of the molecule. They also point out that they expect collagen to persist for much longer than 700,000 years, but in a chemically altered state due to cross-linking reactions and other types of chemical modifications. They state, "Collagen could plausibly be detected at lower concentrations [than 1 percent of the original amounts] in much older material but likely in a diagenetically-altered state and at levels whereby separation from endogenous and exogenous contaminations is much more time-consuming, costly and perhaps applicable only to atypically large taxa that can offer sufficient material for destructive analysis."[22] In other words, diagenesis will produce chemically altered forms of collagen that will persist in animal remains well beyond a million years.

There are no terrestrial taxa more atypically large than dinosaurs. The bodies of these creatures would have contained a huge amount of collagen. The body mass of *Tyrannosaurus rex* ranged from about 5 tons (10,000 lbs.) to about 20 tons (40,000 lbs.). If we assume that about 65 percent of this creature's mass was water and that collagen made up about 25 percent of its dry mass, then a 12-ton *T. rex* would contain about 2,100 lbs. of collagen. Believe it or not! So, 1 percent collagen would equate to 21 lbs. of collagen. Of course, a small piece of the *T. rex* femur would possess a fraction of the total collagen that made up this creature. The point of this rough calculation is to merely point out that there still would be a whole lot of collagen (or collagen-derived material) associated with *T. rex* remains, even after 1 million years have transpired. In this sense, it is really not surprising that paleontologists have been able to detect trace amounts of collagen fragments in dinosaur fossil remains, given how much was there to begin with. The large initial abundances of ovalbumin in dinosaur eggs and shell-binding proteins in mollusk shells also helps explain why paleontologists have been able to detect these materials (or at least fragmented, altered forms of them) in fossilized remains.

---

One final point: I find that when it comes to soft tissue remains, quite a bit of confusion exists about what exactly paleontologists have and have not recovered from fossils. When discussed in popular science outlets, soft tissue discoveries are often referred to as blood vessels, red blood cells, osteocytes, collagen, keratin, chitin, etc. In fact, this practice is even commonplace among paleontologists. Such language gives the impression that scientists have discovered pristine, intact blood vessels, cells, and molecules. But they haven't. They have uncovered materials that have been chemically altered from their original compositions, with some of the original material remaining but most being chemically transformed. They have not isolated intact, unaltered proteins, but fragments. This is how Jack Horner describes these discoveries: "I should point out, also, that calling the fossils red blood cells is a shorthand way of speaking that may be a bit misleading, just like calling a fossilized femur a bone. It is not a bone in the sense that a femur of a recently dead cow is a bone. Minerals in the dinosaur bone have been replaced, chemical changes have occurred. What Mary was seeing, would, if they were real, be remnants of red blood cells with some of the original chemicals remaining, and some of the structure, but with other parts changed forever."[23]

Schweitzer and her fellow researchers did not discover blood vessels, but chemically transformed, chemically cross-linked structures derived from original blood vessels, yet still retaining the original shape. Some of these structures are flexible, but as Schweitzer discovered when surveying dinosaur remains for soft tissue remnants, others are crystallized and inflexible.[24] She did not find original red blood cells and osteocytes, but structures derived from the original cells. It is remarkable that the process preserved structures derived from the nuclei of these cells and even the filopodia of osteocytes, but it did so, most likely through cross-linking reactions.

Paleontologists have not recovered intact proteins such as collagen, keratin, and hemoglobin, but instead, fragments of these molecules.

And that is what Schweitzer and others have found when they sought evidence for proteins in fossil remains. For example, Schweitzer and her team have used antibodies to identify collagen, hemoglobin, and other proteins in dinosaur remains. Antibody binding can be highly specific and highly sensitive, making it a great tool to detect even vanishingly low levels of materials in a sample. But when antibodies bind to molecules, their binding site can only accommodate very small portions of the molecule, called the epitope. This restriction means that the antibodies Schweitzer used in her studies were not necessarily binding to large, intact proteins, but instead attached themselves to small protein fragments. To put it another way, even though the collagen or hemoglobin was highly fragmented, antibodies raised to these proteins would still bind some fragments of the biomolecules.

Schweitzer and her associates also generated amino acid sequences for collagen. But these were sequences not for the entire molecule. Rather, they were for fragments of it. In fact, in one study, Schweitzer's team mapped the amino acid sequences of the collagen fragments isolated from the *T. rex* femur onto molecular models of human and rat collagen fibers.[25] They discovered that the fragments all came from the innermost areas of the fibers, where the strands are packed most closely. These regions are the most protected within the collagen fiber. If the fragments were due to contamination, they should have mapped randomly onto all regions of the collagen fibers. The fact that the fragments clustered to the most protected areas of the fibers makes sense if they were generated from dinosaur collagen.

In another study, researchers from China and Denmark demonstrated that the collagen fragments recovered from dinosaurs possess amino acid compositions and sequences that would make them unusually stable at high temperatures.[26] These two studies shed insight into the preservation of dinosaur collagen, but also yield results expected if these fragments came from ancient remains.

---

Though I have no idea how long it took Josh Chalom to piece together Frank Sinatra's image from 450 Rubik's Cubes, I'm sure he did not accomplish the feat overnight. In like manner, it has taken (and is still taking) paleontologists time to explain how soft tissue remnants persisted in fossil remains for millions of years. While they haven't solved the puzzle quite yet, they have discovered some important clues. And as a result, a beautiful "scientific" picture is emerging.

As I carefully consider the different mechanisms that researchers propose to account for soft tissue survival and as I have thought about this issue myself as a biochemist, I no longer find it difficult to believe that soft tissue fragments can survive for hundreds of millions of years. The existence in fossils of materials derived from blood vessels, cells, and biomolecules is fully compatible with an old-earth view and in no way validates YEC.

Though I find the idea of dinosaur blood and other soft tissues surviving after millions of years as credible, should a layperson believe it or not? This discussion has been admittedly technical, so (in the next few pages) I hope to offer some thoughts by way of recap so you can come to your own conclusion.

CONCLUSION

# Should *You* Believe It or Not?

The *Ripley's Believe It or Not!* Odditoriums throughout the world stand as monuments to the life and accomplishments of Robert Ripley. But the success of Ripley's comic panels, in large measure, is owed to the anonymous work of researcher and fact checker Norbert Pearlroth. Born in Poland, Pearlroth was a polyglot (able to speak multiple languages). In all, Pearlroth was fluent in 14 languages, which made him particularly valuable to Ripley, because he could peruse foreign newspapers in his search for odd facts. Eventually, Pearlroth became Ripley's primary researcher. In that role, he was responsible for delivering 24 oddities a week to the artists and editors who produced the *Ripley's Believe It or Not!* panels. For over 5 decades, Pearlroth spent 10 hours a day, 6 days a week in the New York Public Library researching facts. The library staff estimates that Pearlroth looked through 7,000 books a year. This effort translates into over 350,000 books in the 52 years he worked for *Ripley's Believe It or Not*! Wow!

When claims are hard to believe, it's important to check out the facts. And that is what we attempted to do in this book. We examined the scientific findings reported by Mary Schweitzer and her collaborators (and other teams of paleontologists), who claim to have recovered original soft tissue from the fossilized remains of dinosaurs and other animals. Even though their assertion flies in the face of the standard paradigm for fossilization and, in spite of the protests of a vocal minority of paleontologists, it looks as if their claims are *valid*. Schweitzer and others have compiled a compelling body of evidence for the existence of vessels, cells, and biomolecules in dinosaur bones, feathers, eggshells, and coprolites. Other paleontologists have demonstrated that biomaterials such as quinones, eumelanin, and even chitin can survive in fossil remains for hundreds of millions of years.

These discoveries are forcing a paradigm shift within paleontology. Prior to these advances, almost no one would have thought soft tissues could persist in fossil remains. This breakthrough has ignited a scientific controversy, but it also excites biologists who hope that access to soft tissues will give them new insight into the biology of these long lost creatures.

Young-earth creationists have capitalized on these surprising results, arguing that it is impossible for the fossils to be millions of years old. They maintain that soft tissues should not survive that long. Instead, these materials should readily degrade in a few thousand years. Therefore, the scientifically determined date for the fossils must be in error. In their view, these finds challenge the reliability of radiometric dating methods used to determine the age of these fossils, and along with it Earth's antiquity. Instead, they argue that these breakthrough discoveries provide compelling scientific evidence for a young earth and support the idea that the fossil record results from a recent global (worldwide) flood.

But a little fact checking uncovers reasons to be skeptical of this YEC claim. There are many reasons to be confident in the dates generated by radiometric dating methods for Earth's geological features and the fossil record. The principles and practices that define radiometric techniques are sound. These methods have been validated and cross-correlated with each other and with other dating techniques. As part of the RATE study, YECs have questioned the constancy of the radioactive decay process. If all the radioactive decay rates became accelerated during the creation week and/or at the time of Noah's flood, Earth could have the appearance of being ancient, according to this age-measuring method, while actually being only 6,000 years old. Yet, additional fact checking contradicts this claim, as well.

Bad scientific arguments for the Christian faith negatively impact evangelism. Many skeptics and seekers reject Christianity because they wrongly believe that the Bible teaches that the earth is only 6,000–10,000 years old. This misperception is reinforced by vocal YECs who not only claim that the only valid interpretation of Genesis 1 is the calendar-day view, but also maintain that ample scientific evidence exists for a young earth—such as the recovery of soft tissue remnants in fossils. These questionable scientific claims cause many people to view Christianity as antiscientific, undermining outreach efforts.

Yet, there are other ways to interpret the Genesis 1 creation passage. There is a sound scriptural basis for an old-earth perspective (see Appendix A). In fact, I would argue that a proper reading of biblical passages points to the earth's antiquity. In other words, when the biblical text is interpreted properly, there isn't any conflict between the scientific evidence of the earth's vast age and the creation accounts.

Many people are pleasantly surprised when they learn that scientific discoveries regarding the age of the earth do not undermine Scripture's reliability. In fact, these discoveries support the credibility of the Christian faith.

If the fossils really are millions of years old, then how is it possible for soft tissue to persist for so long? Some fact checking makes it clear that nobody knows the complete answer to that question, but paleontologists are beginning to formulate models to explain soft tissue survival. It appears that a number of distinct mechanisms exist that contribute to soft tissue preservation. While no single mechanism is sufficient to explain why soft tissues survive, their additive effects can, with the "sum being greater than the parts."

If discovery of soft tissue remnants in ancient fossils would have ever made their way onto Norbert Pearlroth's desk, I'm sure he would have felt confident recommending it for a *Ripley's Believe It or Not!* comic panel. Despite appearing unbelievable, the facts check out. Soft tissue remnants *can* survive in fossils for hundreds of millions of years. It may be hard to believe, but it is true! Sometimes scientific facts are stranger than fiction.

APPENDIX A

# A Biblical Case for an Old Earth

What would a place called an "Odditorium" be without a few displays showing off instruments of torture? In *Ripley's Believe It or Not!* Odditorium in Hollywood, one exhibit that always catches my eye features a "heretic beheading axe." As the name implies, this axe was used during the Inquisition of the fifteenth century to remove the heads of heretics—after torturing them for several days.

It's a good thing this practice no longer takes place in the church today. But that doesn't stop Christians from occasionally leveling the charge of heresy against one another. Some YECs believe that any interpretation of Genesis 1 that regards the creation days as anything other than literal, consecutive 24-hour periods of time undermines biblical authority to such an extent, it borders on heresy. Yet, there are very good *biblical* reasons to think that the creation days refer to long, finite periods of time—epochs—not calendar days. This interpretation of the Genesis 1 creation account is called the day-age view.

The day-age view makes it possible to embrace the antiquity of Earth and life on Earth, without compromising the primacy of the creation accounts found in Scripture. Instead of interpreting Genesis 1 in a way that conflicts with science, the day-age view provides a pathway that leads to harmony between science and Christianity.

It is beyond the scope of this work to present a lengthy defense of the day-age view. Instead, I will simply outline some of the biblical and theological reasons why I subscribe to old-earth creationism (OEC) and the day-age view. I urge the reader to keep in mind that my training is in biochemistry. I'm not a biblical scholar, nor a theologian. But I have read countless works of people who are. Therefore, what I present reflects their influence on my thinking. Readers interested in a more detailed and comprehensive defense of the OEC perspective (along with references to the theological literature) should take a look at *A Matter of Days* by astronomer Hugh Ross,[1] *A New Look at an Old Earth*[2] by physicist Don Stoner, *A Biblical Case for an Old Earth*[3] by physicist David Snoke, and *Genesis One and the Age of the Earth* by Hebrew scholar and physicist Rodney Whitefield.[4]

---

One common misperception about the day-age view is that it is a *figurative* reading of Genesis 1. The corollary: only the YEC view treats Genesis 1 *literally*. But this misperception is not reality. Like the YEC view, the day-age view treats Genesis 1 as a divine natural history. Accordingly, the creation days present a chronological ordering of God's creative acts. But instead of providing an exhaustive description of everything God did, the creation days represent "snap shots" of God's handiwork, throughout the course of Earth's and life's history. The Creator did much more than what is mentioned in Genesis 1. This creation account is simply silent about those other acts of creation. Job 37–39, Psalm 104, and Proverbs 8, for example, fill in some missing details.

Like the YEC interpretation, the day-age view is skeptical of aspects of the evolutionary paradigm. Even though the day-age approach regards the fossil record as a real sampling of life's history, it treats that history as divinely orchestrated, with the Creator repeatedly intervening to create new taxa. When it comes to the question of human origins, the day-age and the YEC view have much in common. Both maintain that Scripture teaches that God directly and personally intervened to create human beings in His image. Both interpretations reject human evolution and regard Adam and Eve as historical individuals (the first two humans) who gave rise to all humanity. The primary difference between these two views relates to the meaning of the word "day" in Genesis 1.

As in the English language, the Hebrew word *yom*—translated as day—can refer to: (1) part of a day; (2) the time from sun up to sun down; (3) a calendar day; or (4) a long but finite period of time, an epoch. In other words, a literal reading of Genesis 1 would allow one to translate *yom* as either 24 hours or as an epoch. The day-age view is just as much a literal reading of the Genesis 1 creation account as is the YEC interpretation.

---

From my perspective, several signifiers in the Genesis 1 account indicate that the best reading of *yom* in Genesis 1 is as a long but finite period of time. For example, the first day of creation describes the genesis of the day-night cycle. That is, the first *calendar* day was created on the first *creation* day. This scenario strongly implies that the creation days must be epochs, not 24-hour periods.

Another compelling clue is found in Genesis 2:4. In this passage, the word *yom*—translated as day—refers to the entirety of the creation week. So within the context of the Genesis 1 account, there is at least one instance in which *yom* clearly refers to a time frame that is much longer than 24 hours. This usage opens the door to interpret the other occurrences of *yom* in the same way.

Finally, the omission of the phrase, "And there was evening and there was morning..." from the description of the seventh day suggests that the creation days are epochs, not calendar days. "Evening and morning" are used to connote the conclusion of each of the first six creation days. The fact that this expression is not used for the seventh day implies that this creation day has not ended. Elsewhere in Scripture (Hebrews 4:9, for example), we learn that God's people can still enter into His Sabbath rest. Again, this passage indicates that we are still in the seventh day. In other words, the seventh day of the creation account represents a period of time, not a calendar day. Correspondingly, the other creation days, too, should be understood as periods of time, not 24 hours.

---

Unless the creation days are viewed as epochs, the Genesis 1 and Genesis 2 creation accounts will contradict each other. According to Genesis 1, male and female were made in God's image on the sixth creation day, after God created the land animals. I believe that Genesis 2 amplifies the events of the sixth day, focusing on the creation of Adam and Eve.

The Genesis 2 text teaches that Adam was created from the dust of the earth, before the Garden of Eden existed. It was only after Adam's creation that God planted the garden. According to the text, its growth was through normative means. Adam then worked and maintained the garden. Adam also named the animals. Both of these activities would have taken an extensive period of time. There is no need to maintain and work a garden unless it has been around for a while. And, the process of naming the animals would have required that Adam carefully study the creatures God brought before him. In the Hebrew mindset, the name Adam gave the animals would have had to reflect their dispositions and biological capabilities. It was only after going through a lengthy process that Adam came to recognize that none were suitable for him. It was then that God put him into a deep sleep and created Eve from Adam's side.

All of Adam's experiences described in Genesis 2 would have required a significant amount of time: weeks, months, perhaps years. There is no way that all of the things Adam did could have taken place in 24 hours, let alone in part of a calendar day—before Eve was created.

A YEC interpretation puts Genesis 1 at odds with Genesis 2, undermining the inerrancy of Scripture. On the other hand, a day-age view readily accommodates the amplified account of Adam and Eve's creation in Genesis 2.

---

There are a number of passages throughout the Old and New Testaments that make direct statements about the antiquity of the earth and its features. These statements provide additional biblical support for old-earth creationism, and the day-age view of Genesis 1. A few of those passages are listed below (with emphasis added):

> But they deliberately forgot that *long ago* by God's word the heavens existed and the earth was formed out of water and by water. (2 Peter 3:5)
>
> The LORD brought me forth as the first of his works, before *his deeds of old*; I was appointed from eternity, from the beginning, before the world began. (Proverbs 8:22–23)
>
> To him who rides the *ancient skies* above, who thunders with mighty voice. (Psalm 68:33)
>
> Your father's blessings are greater than the blessings of the *ancient mountains,* than the bounty of the *age-old hills.* (Genesis 49:26)
>
> With the choicest gifts of the *ancient mountains* and the fruitfulness of the *everlasting hills* (Deuteronomy 33:15)
>
> He stood, and shook the earth; he looked, and made the nations tremble. The *ancient mountains* crumbled and the *age-old hills* collapsed. His ways are eternal. (Habakkuk 3:6)
>
> The River Kishon swept them away, the *age-old river*, the River Kishon. (Judges 5:21)

These passages indicate that the Earth is old; the skies are old; the mountains, hills, and rivers are ancient. It is difficult to square these passages with a YEC interpretation of Genesis 1 and the idea that the earth is only 6,000–10,000 years old. On the other hand, these passages align with an old-earth reading of Genesis 1.

---

The biblical and theological case for the day-age interpretation and OEC is much more expansive than I have presented here. There are many compelling biblical and theological reasons to embrace this interpretation. One huge reason is this: the way Christians read Genesis 1 can have a profound impact on their ability to reach others for Christ. A common misconception among Christians (as well as non-Christians) is that the only valid reading of Genesis 1 is the calendar-day view. This causes believers to doubt their faith and leads skeptics and seekers to reject Christianity because they reject the notion that the earth is only 6,000–10,000 years old—an interpretation that is in clear conflict with the scientific record.

However there are many compelling biblical reasons to accept an old-earth perspective, removing a serious objection that skeptics and seekers have to the Christian faith. Hopefully, we can work together to remove unnecessary barriers to faith, making it possible for nonbelievers to entertain the gospel.

APPENDIX B

# The Creation-Evolution Controversy in *Jurassic World*

People are often surprised to learn that I have never seen *Jurassic Park* (or any of its sequels, including *Jurassic World*). Still, that doesn't keep people from asking me if scientists will be able to bring dinosaurs back to life one day. Fortunately, I can answer that question without having seen any of the *Jurassic Park* movies.

Bottom line: Much to the chagrin of sci-fi fans, it is unlikely that scientists will ever be able to resurrect a dinosaur from ancient DNA. This biomolecule cannot be recovered from insects trapped in amber, and it is not stable enough to survive in appreciable levels in dinosaur remains. And even if scientists did have access to a complete dinosaur genome, they would face insurmountable technical hurdles if they tried to convert a dinosaur genome into a living, breathing creature.[1]

Still, there are some researchers—such as the legendary paleontologist Jack Horner—who think we might be able to create dinosaurs using a different approach. Believe it or not, they say it could be possible to transform chickens (and other birds) into dinosaur-like creatures by manipulating their developmental process.[2]

Recent work by researchers from Harvard and Yale Universities moves the scientific community one step closer to creating a "chickenosaurus." These investigators genetically engineered chickens so that they developed snout-like structures instead of beaks, just like that of dinosaurs.[3] For biologists, the significance of this work has little to do with the prospects of bringing dinosaurs back to life, but instead helps them understand how beaks emerged as an evolutionary innovation.

Paleontologists think that birds share an evolutionary history with dinosaurs, currently classifying them as theropods. These scientists argue that shared anatomical features between birds and theropods demonstrate their evolutionary connection. So, too, do the feathered dinosaurs unearthed in the fossil record. And work such as this latest venture by Harvard and Yale scientists reinforces the notion that birds descended from dinosaurs. The research seemingly provides insights into the types of molecular and developmental changes that are responsible for the evolution of birds.

The work also serves as a harbinger of a new approach to paleontology—dubbed reverse evolution. Evolutionary biologists believe they can gain understanding of how biological transformations took place during life's history by experimentally reverting organisms to their ancestral state. Reverse evolution experiments fuse insights from paleontology with those from developmental biology, molecular biology, comparative embryology, and genomics. For the first time, researchers can address questions in evolutionary biology using an experimental strategy.

The investigators from Harvard and Yale chose to study the origin of beaks because they are one of the defining features of modern birds and stand as a key biological innovation in life's history. Birds use beaks for a variety of distinct functions that allow the creatures to exploit a wide range of ecological niches.

By comparing the skeletal anatomy of modern birds with ancient birds (such as *Archaeopteryx*) and theropod dinosaurs found in the fossil record, paleontologists believe that the beak evolved when facial bones called the premaxilla fused and became elongated. Comparisons of the developmental changes that take place in the faces of chickens and emus with those that occur for alligators, turtles, and lizards convinced the researchers that the facial development in dinosaurs was similar to those of modern-day reptiles.

They also noted that during development two proteins (the growth factors dubbed FgF8 and Lef1) were produced by cells in a large region of the face in the developing embryo of chickens, but only in small patches of cells in the developing embryos of reptiles. (Growth factors are proteins and other substances that stimulate cell growth, proliferation, and differentiation.)

Based on these insights, paleontologists reasoned that changes in the expression of the genes that encode these growth factors likely played a key role in the transformation of the snouts of theropods and ancient birds into beaks. To test this idea, they implanted microscopic beads coated with chemicals that inhibit FgF8 and Lef1 into the facial region of chick embryos. To their amazement, the disruption in the growth factors led to chick embryos with snout-like structures rather than beaks, at least with respect to the internal skeletal anatomy. (Externally, the chick embryos had beaks, but internally the skeletal structure looked similar to modern-day reptiles, theropods, and ancient birds.)

At this juncture, the researchers still don't know what the specific genetic changes were that led to altered gene expression of the growth factors. Nor do they believe that they have provided the complete explanation for how beaks evolved from snouts. But they do believe they have gained important insight into key changes that contributed to bird evolution, and that they have demonstrated the power of reverse evolution as a strategy to understand the history of life from an evolutionary framework.

It is remarkable how such a relatively small change can have such a dramatic effect on an organism's anatomy. The Harvard and Yale team (along with other biologists) believes that the dramatic impact caused by subtle changes in gene expression could account for the sudden, saltational changes routinely observed in the fossil record.

---

As impressive as this work is and as compelling as the case for bird evolution *seems* to be, the observations from the fossil record and the results of reverse engineering studies can be readily explained from a creation model perspective.[4] Key to this explanation is the work of Sir Richard Owen, a preeminent biologist who preceded Darwin. In contemporary biology, scientists view shared features possessed by related organisms as evidence of common ancestry. Birds and theropod dinosaurs would be a case in point. But for Owen, shared anatomical features reflected an archetypical design that originated in the Mind of the First Cause.[5] Toward this end, the anatomical features shared by birds and theropods can be understood as reflecting common *design*, as opposed to common *descent*.

Though few biologists take Owen's ideas seriously today, it is important to note that Owen's ideas were not "tried and found wanting." They simply were abandoned in favor of Darwin's theory, which many biologists preferred because it provided a mechanistic explanation for life's history and the origin of biological systems. In fact, Darwin owes a grand debt to Owen's thinking. Darwin coopted the idea of the archetype, but then replaced the canonical blueprint that existed in the Creator's Mind with a hypothetical common ancestor.

This archetypical approach to biology can account for the results of reverse evolution studies, such as the conversion of a bird's beak into a dinosaur snout. Accordingly, the researchers merely "stumbled upon" differences in the developmental program (production of the growth factors FgF8 and Lef1) that affect variations in the archetype—yielding beaks in modern birds, and snouts in modern reptiles and ancient birds and dinosaurs.

From a genetics standpoint, one could argue that the Creator appears to have selected a gene set that can be used to construct a wide range of organisms organized around a single archetype, simply by varying gene expression. An analogy might help. Think of an organism's set of genes like a set of LEGOs. Depending on a child's wishes and imagination, he or she can make numerous structures from the same set of bricks. The builder makes use of the same pieces, but employs them differently to create a variety of structures. In like manner, a Creator could produce different organisms—ones with beaks and ones with snouts—from the same set of genes, by deploying those genes in different ways.

In effect, what I have described is a common design principle. Computer engineers produce computers with fixed hardware that can be programmed for a bewildering array of functions. Even human languages rely on this principle. A relatively small set of words can be used to communicate a seemingly endless number of ideas and concepts.

From a creation model standpoint, the researchers from Harvard and Yale Universities did not reverse the evolutionary process. They unwittingly reverse engineered a snout from a beak based on design principles.

The amount of knowledge and insight into the developmental process that was required to carry out these (and other) reverse evolution experiments adds further credibility to a creation model interpretation of this work. Equally impressive was the care required to alter the influence of the two growth factors. In other words, to transform a beak into a snout required coordinated, coherent, and precisely localized changes in growth factor levels. That is, the researchers intelligently designed the conversion of the chicken's beak into snout-like structures. This being so, is it plausible to think that unguided, historically contingent processes could carry out such transformations when small changes can have profound effects on an organism's anatomy? It seems that the best the evolutionary process could achieve would be the generation of "monsters" with little hope of survival. Why? Because evolutionary mechanisms can only change gene expression patterns in a random, haphazard manner.

I would contend that the coherent, precisely coordinated genetic changes needed to generate one biological system from another bespeaks not undirected evolutionary mechanisms, but a Creator's handiwork, as the explanation for life's history.

APPENDIX C

# Dinosaur Genome Size Estimates: *Lagerstätten* of Design

A hard-to-believe paper was published in *Nature* in 2007 by researchers from the UK. In this report, the scientific investigators describe the first-ever estimates of genome sizes for dinosaurs—even though there was no DNA left behind in their remains![1]

Provocatively, a byproduct of this work highlights another important function for so-called junk DNA—providing further evidence that the genomes of creatures such as dinosaurs reflect the elegant handiwork of a Creator.

The research team used an ingenious approach to estimate dinosaur genome size. Workers took advantage of the relationship between cell size and genome size. It turns out that larger cells tend to have more DNA in the nucleus than smaller cells.

Cell biologists believe that in addition to housing genetic information, DNA plays a skeletal role in the nucleus; the more DNA in the nucleus, the larger the volume of this organelle because of the space taken up by DNA. The ratio of the nuclear volume-to-cell volume is a critical parameter. To accommodate reasonable growth rates, large cells need to have a large nucleus. A large nucleus has a large surface area. This characteristic allows for rapid transport of messenger RNA from inside the nucleus to the cytoplasm, where it can direct protein synthesis. Transport of messenger RNA occurs through nuclear pores. A large nucleus has more pores than a small one because of its greater surface area. Large cells require a much higher rate of protein synthesis than do smaller cells, in order to have enough subcellular and biochemical components to sustain their activity. This requirement necessitates a sufficiently rapid transport of messenger RNA to the cytoplasm.

Researchers have noted that when dinosaur bones are sliced, the outlines of bone cells, or osteoclasts, are evident. The team used this observation to estimate osteoclast size for a number of dinosaurs from fossilized bone slices. They then established a calibration for genome size by correlating DNA content to osteoclast size for a number of living vertebrates (amphibians, reptiles, birds, and mammals).

Investigators discovered that dinosaur genome sizes cluster into two groupings that correspond to theropod and nontheropod dinosaurs. They found that theropod dinosaurs had a much smaller genome size, on average, than did their nontheropod counterparts.

The scientists rationalized this find by noting that other evidence suggests theropods had a much higher metabolic activity rate than nontheropod dinosaurs. Smaller cell sizes would benefit creatures with high metabolic activity. Smaller red blood cells have a higher surface-to-volume ratio and, consequently, a more efficient exchange of gases (like oxygen and carbon dioxide) with surrounding tissues.

Researchers speculated that the amount of noncoding DNA found in their genomes dictated the difference in cell sizes for the two groups of dinosaurs. Presumably, the larger genomes of nontheropods contained more noncoding DNA than theropod genomes.

In broad terms, the genomes of organisms consist of coding DNA (which harbors genes) and noncoding DNA. For most organisms, noncoding DNA comprises a vast majority of their genomes. Over 95 percent of the human genome, for example, consists of noncoding DNA.

Evolutionary biologists have traditionally thought of noncoding DNA as junk—the product of random biochemical events. They consider the existence of junk DNA as one of the most potent pieces of evidence for biological evolution. According to this view, junk DNA results when undirected biochemical processes and random chemical and physical events transform a functional DNA segment into a useless molecular artifact. Junk pieces of DNA remain part of an organism's genome solely because of its attachment to functional DNA. In this way, junk DNA persists from generation to generation, and is thought to be a vestige of an evolutionary history. Skeptics ask, "Why would a Creator purposely introduce nonfunctional, junk DNA into the genomes of organisms?"

As a creationist, I would ask, "Why would evolutionary mechanisms tolerate such large amounts of nonfunctional DNA within genomes of organisms?" It would be energetically expensive for an organism to maintain nonessential DNA sequences. The presence of large amounts of noncoding DNA in genomes, alone, suggests this DNA has function. And the estimates of dinosaur genome sizes affirm the notion that noncoding DNA is not junk at all. Rather, some of these DNA sequences play a critical role as a nucleoskeleton.

Studies like these help explain why the Creator purposely introduced noncoding DNA into the genomes of organisms. Junk DNA is a veritable lagerstätte (storage place) of design.

# Glossary

**A**

**N-acetylglucosamine:** A carbohydrate that serves as a key component of chitin

**actin:** A globular protein that forms filaments that are part of the cell's cytoskeleton

**alanine:** An amino acid that occurs naturally in proteins

**alkaline phosphatase:** An enzyme that plays a key role in the formation of the mineral phase of bone

**amino acids:** A class of organic compounds that serve as subunits and are linked together to form proteins

**anhydrous:** A system or material that is completely dry, containing no water

**antibodies:** A Y-shaped protein that plays a role in the immune response by binding to pathogens and other foreign materials, flagging them for removal

**antimicrobial activity:** Chemical or physical processes that inhibit the growth of bacteria, typically by bringing about cell death

**aragonite:** A mineral made from calcium carbonate

**aromatic bond:** A highly stable chemical bond found in certain ringed organic compounds. Aromatic bonds typically form when the bonds within the ring alternate between single and double bonds. Aromatic rings form readily and, once formed, are extremely difficult to break.

**Arrhenius equation:** A mathematical relationship that describes the temperature dependence of the reaction rate for chemical transformations

**arthropods:** A phylum characterized by exoskeletons and segmented bodies. This group includes insects, spiders, and crustaceans.

**B**

**basement membrane:** The part of the extracellular matrix that anchors cells in tissues

**biliverdin:** A pigment compound found in bile. Biliverdin is a breakdown product of heme.

**bioapatite:** The mineral component of bone that is made up of calcium phosphate

**biodegradation:** The breakdown of any material mediated by the activity of microorganisms

**biofilm:** A thin film of mucous created by microorganisms, such as bacteria. The mucous film houses the microorganisms that produced it.

**biomolecules:** Molecules, such as DNA, proteins, lipids, and carbohydrates, that occur naturally in living organisms

## C

**calcite:** A mineral composed of a crystalline form of calcium carbonate

**calcium carbonate:** An inorganic compound made from the calcium and carbonate ions. It occurs widely in limestone, chalk, and marble.

**calcium phosphate:** An inorganic compound made from calcium and phosphate ions. This compound is the main mineral component of bone.

**capillary:** A small blood vessel that joins arteries to veins

**carbohydrate:** A naturally occurring class of molecules that includes sucrose, starch, and cellulose

**carbon-14:** The radioactive isotope of carbon used in carbon dating techniques

**carbonaceous material:** Any material that is rich in carbon

**cellulose:** A polysaccharide made up of glucose. Cellulose forms the main constituents of plants' cell walls and is used to manufacture paper and textiles.

**chelating agent:** A compound that binds and sequesters positively charged metal ions

**chitin:** A polysaccharide made up of N-acetylglucosamine that forms the exoskeleton of arthropods

**coding DNA:** Sequences of DNA that contain the information the cell's machinery needs to make proteins

**collagen:** A class of elongated fibrous proteins that serve as a major constituent of connective tissue

**collagenase:** Enzymes that degrade or break down the protein collagen

**constants of physics:** A numerical quantity that describes some aspect of the universe. This quantity is universal and unchanging with respect to time.

**coprolite:** Fossilized dung

**cosmic radiation:** High-energy radiation that originates outside of the solar system, but is of unknown origin

**covalent bond:** A type of chemical bond between two atoms that involves sharing electrons between nuclei

**cross correlation:** A process that involves comparing the results of two different analytical methods as a way to determine the reliability of one or both methods

**cross-links:** Chemical bonds that link one polymer chain to another. These bonds can be either covalent or ionic bonds.

**cuticle:** In its general use, this term refers to any organic material that serves as a tough, flexible covering for living creatures. For invertebrates, the cuticle is a multilayered sheet that surrounds the skin to form the exoskeleton. In arthropods, chitin is the major component of the cuticle's individual sheets.

**cysteine:** An amino acid that naturally occurs in proteins

## D

**d-:** A designation given to a compound that can exist in one of two forms that are mirror images of each other

**demineralization:** Any process that removes minerals from biological tissues, in materials such as bones

**diagenesis:** Changes that take place in sediments, sedimentary rocks, and fossils due to the high temperatures and pressures associated with geological processes

**disulfide bridge:** A cross-link that occurs in proteins that are formed by the covalent bond between two cysteine amino acids

**DNA:** The molecule that carries genetic information

## E

**elastin:** A protein found in connective tissue. Elastin imparts to tissue the capacity to resume its original shape after being deformed.

**electron delocalization:** Refers to electrons in a molecule that are not associated with a single atom or covalent bond

**electrostatic bond:** A chemical bond formed between ions (positively and negatively charged atoms)

**endogenous:** Substances and processes that are inherent parts of an organism

**endothelial cell:** A specialized type of cell that lines the interior of blood vessels

**energy of activation:** The minimum energy that must be available for chemical compounds before they undergo a chemical reaction

**entropy:** A thermodynamic quantity that describes the energy quality of a system. Systems characterized by high entropy have low energy quality and, hence, a low capacity to do work.

**environmental enzyme:** An enzyme released into the environment by an organism through secretion, or released upon the organism's death

**enzyme:** A general class of proteins that serve as catalysts that accelerate chemical reactions

**ethylenediaminetetraacetic acid (EDTA):** This compound is a commonly used chelating agent that can sequester calcium. Hence, EDTA can be used to demineralize bone tissue.

**eumelanin:** A member of a class of naturally occurring pigmented compounds. The two types of eumelanin are called black and brown.

**exogenous:** Substances and processes originating outside the organism

**extracellular matrix:** A molecular film, secreted by cells, that serves as a site for cell adhesion and provides the means for cell-to-cell communication, thus providing structural and biochemical support

F

**feldspar:** A collection of aluminum silicate minerals that make up about 60 percent of Earth's crust

**femur:** The large, long bone of the upper leg (thigh)

**ferritin:** A group of large iron-carrying proteins

**filopodia:** A slender projection that extends from the surface of some cells

**fluvial deposits:** Geological deposits formed by rivers and streams

**formaldehyde:** A small, naturally occurring organic compound that, when dissolved in water, forms formalin. Formaldehyde is often used as a disinfecting agent and an embalming compound.

G

**gastropods:** A large group of organisms, including slugs and snails, within the phylum Mollusca

**genetic engineering:** The use of laboratory techniques to alter an organism's genetic makeup. Genetic engineering can include transferring genes from one organism to another.

**genome:** An organism's complete set of genetic information housed in the DNA found in the nucleus

**glycine:** A naturally occurring amino acid found in proteins. Glycine occurs in high levels in collagen.

**graphite:** A form of carbon where six atoms are organized into hexagonal rings. The rings are then fused together into two-dimensional sheets and stacked on top of each other. Graphite is the most thermodynamically stable form of carbon.

**growth factor:** A naturally occurring substance (either a steroid hormone or protein) that can stimulate the growth and differentiation of cells

## H

**heme:** A complex structure consisting of a central iron atom surrounded by a porphyrin ring. The heme associated with hemoglobin serves as the oxygen-binding site.

**hemoglobin:** A protein found in red blood cells that binds and transports oxygen to the tissues of the body

**histone:** Proteins that bind to DNA and organize the biomolecule into nucleosomes, which are the foundational structures of chromosomes

**hydrogen bond:** A weak electrostatic interaction between molecules involving a hydrogen atom bound to oxygen, nitrogen, or fluorine

**hydrolysis:** A type of chemical reaction in which a water molecule cleaves or breaks a chemical bond

**hydrolytic enzymes:** An enzyme that catalyzes the breaking of a chemical bond via a hydrolysis reaction

**hydrophobic:** In chemistry, this term refers to a substance's physical property that allows it to repel water.

**hydroxyapatite:** The mineral component of bone made up of calcium phosphate. By weight, roughly 70 percent of human bone is hydroxyapatite.

**hydroxylysine:** A naturally occurring amino acid that plays a key role in cross-linking collagen chains

**hydroxyproline:** A naturally occurring amino acid found at high levels in collagen. It allows collagen chains to turn sharply so that they can form a triple helix; it also stabilizes the triple helix.

## I

**ion:** A charged atom that forms either by losing (positively charged) or gaining (negatively charged) electrons

**isotopes:** Differing forms of an element that possess an identical proton count and atomic number, but have a different neutron count within their nuclei and a different mass number

## J

**junk DNA:** Any DNA sequence in an organism's genome that lacks function

## K

**keratin:** A family of fibrous proteins that are major components of hair, nails, claws, and feathers

**kinetically controlled chemical reaction:** A chemical reaction that proceeds more rapidly than thermodynamically controlled chemical reactions and yields more unstable products due to a lower energy of activation

**kinetics:** The study of chemical reaction rates

## L

**l-:** A designation given to a compound that can exist in one of two forms that are mirror images of each other

***Lagerstätten*:** German for "storage places." In paleontology, it refers to sedimentary deposits that contain large numbers of exceptionally well-preserved fossils.

**laminin:** A very large protein found in the basement membrane of tissues. The basement membrane is part of the extracellular matrix that anchors cells in tissues.

**leucine:** An amino acid that occurs naturally in proteins

**lipids:** A structurally diverse class of biomolecules, including fats and cholesterol, that are water insoluble

**lysine:** A positively charged amino acid that occurs naturally in proteins

## M

**melanin:** A group of naturally occurring dark pigments found in hair, skin, nails, and feathers

**melanosome:** An organelle where melanin is made and stored found in skin cells and cells of the retinal epithelium

**messenger RNA:** A molecule produced by the cell's machinery that contains a copy of the genetic information harbored in DNA. Messenger RNA makes its way to the ribosome where it directs the production of proteins.

**metabolically active:** Cells or tissues that are actively carrying out biochemical and cellular processes

**meteorite:** An object made of metal or stone that fell to Earth as a remnant of a meteor

**mineralogy:** The study of the materials and properties that make up a mineral

**moiety:** In chemistry, this term refers to a part of a molecule (sometimes called a functional group).

**muscle fiber:** Striated fibers that form when muscle cells fuse together in a linear fashion

## N

**nitrogen-14:** A stable isotope of nitrogen. Nitrogen-14 is one of the end products of the radioactive decay of carbon-14.

**non-heme iron protein:** A protein that contains an iron cofactor, but lacks a heme moiety. Typically, in non-heme iron proteins, the iron binds to the active site by forming a complex with the side chains of cysteine amino acids.

**nuclear pore:** A dynamic opening in the nucleus' membrane that plays a role in transporting materials in and out of the nucleus

**nucleoskeleton:** A role proposed for DNA, where this biomolecule provides the structural framework for the nucleus, thus establishing its volume

**nucleus:** A large spherical organelle that harbors genetic material and is found in eukaryotic (or complex) cells

## O

**organelle:** A specialized structure, such as the nucleus or melanosome, inside eukaryotic (complex) cells that performs a specific function

**organic compound:** Any chemical compound, whether naturally occurring or man-made, containing carbon

**osteoblast:** A cell that plays a role in bone formation

**osteocalcin:** A protein, produced by osteoblasts, that plays a role in bone formation and in regulating the calcium ions needed to form and maintain the mineral phase of bone

**osteocyte:** A bone cell that plays a role in the maintenance of bone tissue

**ovalbumin:** A protein that is the major component of egg whites

**oxidation:** A chemical reaction where oxygen is added to another substance. Also, this term can refer to chemical reactions where the reactants lose electrons.

## P

**paleontologist:** A scientist who studies the history on Earth through the fossil record

**pH:** A numerical scale used to quantitatively measure a solution's acidity or alkalinity

**pheomelanin:** A compound, red in color, that belongs to a general class of naturally occurring pigments referred to as melanin

**PHEX:** A protein associated with cell membranes that plays a role in bone and teeth formation

**photoprotection:** A collection of mechanisms that protect organisms from the harmful effects of visible light and UV radiation

**polymer:** A large molecule composed of repeated subunits (called monomers). DNA and proteins are examples of polymers found in cells.

**polysaccharide:** Carbohydrate polymers made up of sugar subunits. Cellulose and chitin are two examples of polysaccharides.

**porphyrin:** A class of naturally occurring large-ringed compounds, such as heme, formed by linking together four pyrrole subunits

**premaxilla:** A pair of cranial bones at the tip of the upper jaw, which may or may not bear teeth. In some animals, premaxilla bones are fused.

**proline:** A naturally occurring amino acid found at high levels in collagen

**protein:** Polymers comprised of amino acids that play a wide range of disparate roles in living systems

**protein synthesis:** The collection of metabolic reactions taking place inside the cell that form proteins

**protoporphyrin:** A class of chemical compounds structurally related to porphyrins. Protoporphyrins are employed as pigments in birds' eggs.

**pyroxene:** Rock-forming silicate minerals

**pyrrole:** An organic compound made up of a five-membered ring. Pyrrole is a subunit of porphyrins, such as heme.

## Q

**quinones:** A class of organic compounds that possess one or more aromatic rings

## R

**racemization:** The chemical conversion of a pure chiral compound to a 50:50 mixture of both chiral forms

**radioactive decay:** The process by which an unstable radioactive isotope of an atom emits subatomic particles and high energy to transform into a smaller, stable atom

**radioactive isotopes:** An atom with an unstable nuclear configuration that will eventually undergo radioactive decay

**radioactivity:** The emission of high energy as a result of radioactive decay

**radiocarbon:** The radioactive form of carbon, such as carbon-14

**radiometric dating:** The use of radioactive decay processes and the ratios of radioactive isotopes to determine the age of materials, objects, or geological layers

**radioprotection:** Protection against high-energy radiation produced from radioactive decay

**reverse evolution:** A sub-discipline within evolutionary biology that strives to understand evolutionary transformations by attempting to revert an organism to its ancestral state through genetic engineering

## S

**saltational change:** In biology, this term refers to a sudden, non-gradual change in an organism from generation to generation.

**shell-binding proteins:** Proteins that interact with calcite to form the shells of creatures such as mollusks

**spectroscopic technique:** Any method that seeks to gain information about a material by monitoring its interaction with electromagnetic radiation

**spontaneous chemical reaction:** A chemical reaction that is thermodynamically permissible

**stable isotope:** An isotope of an atom that will not undergo radioactive decay

**starch:** A polysaccharide formed from glucose that serves as a source of stored chemical energy, making it an ideal food staple for humans and animals

**subunit:** This term can refer to a molecule that binds to other molecules to form polymer chains. It can also refer to a single protein chain that, when folded, interacts with other proteins to form a complex.

**synchrotron radiation:** Electromagnetic radiation generated when charged subatomic particles are accelerated

## T

**thermodynamically controlled chemical reaction:** A chemical reaction that proceeds to a greater extent than other competing chemical processes because it yields more stable products

**thermodynamics:** The science of studying heat and the conversions between the different forms of energy

**triple helix:** In biochemistry, a helix formed by intertwining three protein chains together in a helical manner. The triple helix characterizes the foundational structure of collagen.

**tubulin:** One of two different proteins that are used to form microtubules, which are part of the cell's cytoskeleton

## V

**validation:** The process of determining if a scientific method produces reliable results

## Z

**zircon:** A mineral made up of silica and zirconia

# Notes

## Introduction

1. For example, see Brian Thomas, "A Review of Original Tissue Fossils and Their Age Implications," in *Proceedings of the Seventh International Conference on Creationism* (Pittsburgh, PA: Creation Science Fellowship, 2013); Brian Thomas, "Original Biomaterials in Fossils," *Creation Research Society Quarterly* 51 (Spring 2015): 234–47.
2. Mark Hollis Armitage and Kevin Lee Anderson, "Soft Sheets of Fibrillar Bone from a Fossil of the Supraorbital Horn of the Dinosaur *Triceratops horridus*," *Acta Histochemica* 115 (July 2013): 603–8, doi:10.1016/j.acthis.2013.01.001.
3. Brian Thomas and Vance Nelson, "Radiocarbon in Dinosaur and Other Fossils," *Creation Research Society Quarterly* 51 (Spring 2015): 299–311.

## Chapter 1

1. Jeff Hecht, "First Evidence That Dinosaurs Laid Colorful Blue-Green Eggs," *New Scientist*, May 21, 2015, https://www.newscientist.com/article/dn27573-first-evidence-that-dinosaurs-laid-colourful-blue-green-eggs/; Jasmina Wiemann et al., "The Blue-Green Eggs of Dinosaurs: How Fossil Metabolites Provide Insights into the Evolution of Bird Reproduction," *PeerJ PrePrints*, submitted on May 15, 2015, doi:10.7287/peerj.preprints.1080v1,
2. M. H. Schweitzer et al., "Molecular Preservation in Late Cretaceous Sauropod Dinosaur Eggshells," *Proceedings of the Royal Society B* 272 (April 2005): 775–84, doi:10.1098/rspb.2004.2876.
3. Tim Stafford, *The Adam Quest: Eleven Scientists Who Held on to a Strong Faith while Wrestling with the Mystery of Human Origins* (Nashville: Nelson Books, 2013), 103–15.
4. For a detailed recounting of this story from Mary Schweitzer's perspective, see Stafford, *The Adam Quest,* 103–15. For a recounting of the story from Jack Horner's perspective (Schweitzer's PhD research advisor), see Jack Horner and James Gorman, *How to Build a Dinosaur: The New Science of Reverse Engineering* (New York: PLUME, 2010), 57–113.
5. Mary Higby Schweitzer et al., "Preservation of Biomolecules in

Cancellous Bone of *Tyrannosaurus rex*," *Journal of Vertebrate Paleontology* 17 (June 1997): 349–59, doi:10.1080/02724634.1997.10010979 ; Mary H. Schweitzer et al., "Heme Compounds in Dinosaur Trabecular Bone," *Proceedings of the National Academy of Sciences, USA* 94 (June 1997): 6291–96.

6. Mary H. Schweitzer, "Blood from Stone: How Fossils Can Preserve Soft Tissue," *Scientific American*, December 2010, 62–69, http://www.scientificamerican.com/article/blood-from-stone/.
7. Mary H. Schweitzer et al., "Soft-Tissue Vessels and Cellular Preservation in *Tyrannosaurus rex*," *Science* 307 (March 2005): 1952–55, doi: 10.1126/science.1108397.
8. Mary Higby Schweitzer et al., "Analyses of Soft Tissue from *Tyrannosaurus rex* Suggest the Presence of Protein," *Science* 316 (April 2007): 277–80, doi:10.1126/science.1138709; John M. Asara et al., "Protein Sequences from Mastodon and *Tyrannosaurus rex* Revealed by Mass Spectrometry," *Science* 316 (April 2007): 280–85, doi:10.1126/science.1137614; Chris L. Organ et al., "Molecular Phylogenetics of Mastodon and *Tyrannosaurus rex*," *Science* 320 (April 2008): 499.
9. Mary Higby Schweitzer et al., "Molecular Analysis of Dinosaur Osteocytes Support the Presence of Endogenous Molecules," *Bone* 52 (January 2013): 414–23.
10. Mary H. Schweitzer et al., "Biomolecular Characterization and Protein Sequences of the Campanian Hadrosaur *B. canadensis*," *Science* 324 (May 2009): 626–31, doi:10.1126/science.1165069.
11. Robert R. Reisz et al., "Embryology of Early Jurassic Dinosaurs from China with Evidence of Preserved Organic Remains," *Nature* 496 (April 2013): 210–14, doi:10.1038/nature11978.
12. Mary Higby Schweitzer, Jennifer L. Wittmeyer, and John R. Horner, "Soft Tissue and Cellular Preservation in Vertebrate Skeletal Elements from the Cretaceous to the Present," *Proceedings of the Royal Society B* 274 (January 2006): 183–97, doi:10.1098/rspb.2006.3705.
13. Sergio Bertazzo et al., "Fibres and Cellular Structures Preserved in 75-Million-Year-Old Dinosaur Specimens," *Nature Communications* 6 (June 2015): id. 7352, doi:10.1038/ncomms8352.
14. Mark Hollis Armitage and Kevin Lee Anderson, "Soft Sheets of Fibrillar Bone from a Fossil of the Supraorbital Horn of the Dinosaur *Triceratops horridus*," *Acta Histochemica* 115 (July 2013): 603–8, doi:10.1016/j.acthis.2013.01.001.
15. M. H. Schweitzer et al., "Beta-Keratin Specific Immunological Reactivity in Feather-Like Structures of the Cretaceous Alvarezsaurid, *Shuvuuia deserti*," *Journal of Experimental Zoology* 285 (August 1999):

146–57.

16. Mary H. Schweitzer et al., "Keratin Immunoreactivity in the Late Cretaceous Bird *Rahonavis ostromi*," *Journal of Vertebrate Paleontology* 19, no. 4 (1999): 712–22, doi:10.1080/02724634.1999.10011183.
17. Johan Lindgren et al., "Molecular Composition and Ultrastructure of Jurassic Paravian Feathers," *Scientific Reports* 5 (August 2015): id. 13520, doi:10.1038/srep13520.
18. Karen Chin et al., "Remarkable Preservation of Undigested Muscle Tissue within a Late Cretaceous Tyrannosaurid Coprolite from Alberta, Canada," *Palaios* 18 (June 2003): 286–94.
19. J. R. Nance et al., "Preserved Macroscopic Polymeric Sheets of Shell-Binding Protein in the Middle Miocene (8 to 18 Ma) Gastropod *Ecphora*," *Geochemical Perspective Letters* 1 (January 2015): 1–9, doi:10.7185/geochemlet.1501.
20. Dale E. Greenwalt et al., "Hemoglobin-Derived Porphyrins Preserved in a Middle Eocene Blood-Engorged Mosquito," *Proceedings of the National Academy of Sciences, USA* 110 (November 2013): 18496–500, doi:10.1073/pnas.1310885110.
21. N. P. Edwards et al., "Infrared Mapping Resolves Soft Tissue Preservation in 50 Million Year-Old Reptile Skin," *Proceedings of the Royal Society B* 278 (November 2011): 3209–18, doi:10.1098/rspb.2011.0135.
22. Keely Glass et al., "Direct Chemical Evidence for Eumelanin Pigment from the Jurassic Period," *Proceedings of the National Academy of Sciences, USA* 109 (June 2012): 10218–23, doi:10.1073/pnas.1118448109.
23. Christina E. O'Malley, William I. Ausich, and Yu-Ping Chin, "Isolation and Characterization of the Earliest Taxon-Specific Organic Molecules (*Mississippian, Crinoidea*)," *Geology* 41 (March 2013): 347–50, doi:10.1130/G33792.1.
24. B. Artur Stankiewicz et al., "Preservation of Chitin in 25-Million-Year-Old Fossils," *Science* 276 (June 1997): 1541–43, doi:10.1126/science.276.5318.1541.
25. Patricia G. Weaver et al., "Characterization of Organics Consistent with β-Chitin Preserved in the Late Eocene Cuttlefish *Mississaepia mississippiensis*," *PLoS ONE* 6 (November 2011): e28195, doi:10.1371/journal.pone.0028195.
26. George D. Cody et al., "Molecular Signature of Chitin-Protein Complex in Paleozoic Arthropods," *Geology* 39 (March 2011): 255–58, doi:10.1130/G31648.1.
27. H. Ehrlich et al., "Discovery of 505-Million-Year Old Chitin in the Basal Demosponge *Vauxia gracilenta*," *Scientific Reports* (December 13, 2013): id. 3497, doi:10.1038/srep03497.

28. Schweitzer, "Blood from Stone," 64.
29. Ibid., 67.
30. Thomas G. Kaye, Gary Gaugler, and Zbigniew Sawlowicz, "Dinosaurian Soft Tissues Interpreted as Bacterial Biofilms," *PLoS ONE* 7 (July 2008), doi: 10.1371/journal.pone.0002808.
31. Marshall Bern, Brett S. Phinney, and David Goldberg, "Reanalysis of *Tyrannosaurus rex* Mass Spectra," *Journal of Proteome Research* 8 (September 2009): 4328–32, doi:10.1021/pr900349r.
32. Ibid.
33. Timothy P. Cleland et al., "Mass Spectrometry and Antibody-Based Characterization of Blood Vessels from *Brachylophosaurus canadensis*," *Journal of Proteome Research* 14 (December 2015): 5252–62, doi:10.1021/acs.jproteome.5b00675.
34. Schweitzer, "Blood from Stone," 67.


**Chapter 2**


1. Kevin Anderson, "The iDINO Project," *Creation Research Society Quarterly* 51 (Spring 2015): 232.
2. Brian Thomas, "A Review of Original Tissue in Fossils and Their Age Implications," in *Proceedings of the Seventh International Conference on Creationism* (Pittsburgh, PA: Creation Science Fellowship, 2013); Brian Thomas, "Original Biomaterials in Fossils," *Creation Research Society Quarterly* 51 (Spring 2015): 234–47.
3. Anderson, "The iDINO Project," 233.
4. Thomas G. Kaye, Gary Gaugler, and Zbigniew Sawlowicz, "Dinosaurian Soft Tissues Interpreted as Bacterial Biofilms," *PLoS ONE* 7 (July 2008), doi: 10.1371/journal.pone.0002808.
5. For example, see B. Klont and J. M. ten Cate, "Susceptibility of the Collagenous Matrix from Bovine Incisor Roots to Proteolysis after *in vitro* Lesion Formation," *Caries Research* 25, no. 1 (1991): 46–50, doi:10.1159/000261341; Matthew J. Collins et al., "A Basic Mathematical Simulation of the Chemical Degradation of Ancient Collagen," *Journal of Archeological Science* 22 (March 1995): 175–83; M. J. Collins et al., "The Survival of Organic Matter in Bone: A Review," *Archaeometry* 44 (August 2002): 383–94, doi: 10.1111/1475-4754.t01-1-00071.
6. Thomas, "A Review of Original Tissue"; Thomas, "Original Biomaterials in Fossils," 234–47.
7. Mike Buckley and Matthew James Collins, "Collagen Survival and Its Use for Species Identification in Holocene-Lower Pleistocene Bone Fragments from British Archeological and Paleontological Sites," *Antiqua* 1, no. 1 (2011): e1, doi:10.4081/antiqua.2011.e1.

8. Mike Buckley et al., "Comment on 'Protein Sequences from Mastodon and *Tyrannosaurus rex* Revealed by Mass Spectrometry,'" *Science* 319 (January 2008): 33c, doi:10.1126/science.1147046.
9. Anderson, "The iDINO Project," 228–33.
10. Barry Yeoman, "Schweitzer's Dangerous Discovery," *Barry Yeoman* (blog), published April 1, 2006, http://barryyeoman.com/2006/04/schweitzers-dangerous-discovery/. This article was originally published in *Discover* (April 2006).
11. Thomas, "A Review of Original Tissue."
12. Brian Thomas and Vance Nelson, "Radiocarbon in Dinosaur and Other Fossils," *Creation Research Society Quarterly* 51 (Spring 2015): 299–311.
13. Ibid., 299.
14. Anderson, "The iDINO Project," 232.

**Chapter 3**

1. "T. Rex Blood?: Expert Q&A," *Nova ScienceNow*, July 31, 2007, http://pbs.org/wgbh/nova/nature/schweitzer-qa.html.
2. Emily Ruppel and Mary Schweitzer, "Not So Dry Bones: An Interview with Mary Schweitzer," BioLogos, published July 21, 2014, http://biologos.org/blog/not-so-dry-bones-an-interview-with-mary-schweitzer.
3. Hugh Ross, *A Matter of Days: Resolving a Creation Controversy*, 2nd exp. ed. (Covina, CA: RTB Press, 2015), 145–219.
4. Don Stoner, *A New Look at an Old Earth: Resolving the Conflict between the Bible and Science* (Eugene, OR: Harvest House, 1997), 57–116.
5. David Snoke, *A Biblical Case for an Old Earth* (Grand Rapids: Baker, 2006), 24–46.
6. Carol Hill, Gregg Davidson, Wayne Ranney, and Tim Helble, eds., *The Grand Canyon, Monument to an Ancient Earth: Can Noah's Flood Explain the Grand Canyon?*, (Grand Rapids, MI: Kregel Publishers, 2016).
7. G. Brent Dalrymple, *The Age of the Earth* (Stanford, CA: Stanford University Press, 1991).
8. G. Brent Dalrymple, *Ancient Earth, Ancient Skies: The Age of the Earth and Its Cosmic Surroundings* (Stanford, CA: Stanford University Press, 2004).
9. G. Brent Dalrymple, "Radiometric Dating Does Work!," *Reports of the National Center for Science Education* 20 (May–June 2000): 14–19.
10. Davis A. Young and Ralph F. Stearly, *The Bible, Rocks and Time: The*

*Geological Evidence for the Age of the Earth* (Downers Grove, IL: Intervarsity Press, 2008), 388–446.
11. Roger Weins, "Radiometric Dating: A Christian Perspective," *Science in Christian Perspective*, revised 2002, http://asa3.org/ASA/resources/Wiens.html.
12. Dalrymple, *Age of the Earth*, 86–90.
13. Dalrymple, *Ancient Earth, Ancient Skies*, 54.
14. Steven A. Austin, "Excess Argon within Mineral Concentrates from the New Dacite Lava Dome at Mount St. Helens Volcano," *Creation Ex Nihilo Technical Journal* 10, part 3 (1996): 335–43.
15. Weins, "Radiometric Dating," http://asa3.org/ASA/resources/Wiens.html.
16. Dalrymple, "Radiometric Dating Does Work!," 14–19.
17. P. R. Renne, et al., "$^{40}Ar/^{39}Ar$ Dating into the Historical Realm: Calibration against Pliny the Younger," *Science* 277 (August 1997): 1279–80.
18. Weins, "Radiometric Dating," http://asa3.org/ASA/resources/Wiens.html.
19. Dalrymple, *The Age of the Earth*, 257–304.
20. Dalrymple, *Ancient Earth, Ancient Skies*.
21. Kenneth R. Miller, *Finding Darwin's God: A Scientist's Search for Common Ground between God and Evolution* (New York: Cliff Street Books, 1999), 68.
22. Dalrymple, *Ancient Earth, Ancient Skies*, 58.
23. John D. Morris, "RATE Group to Release Book," *Acts and Facts* 29, no. 8 (2000), http://www.icr.org/article/rate-group-release-book.
24. Larry Vardiman, Andrew Snelling, and Eugene F. Chaffin, eds., *Radiosotopes and the Age of the Earth: A Young-Earth Creationist Research Initiative* (El Cajon, CA: Institute for Creation Research, 2000), 3.
25. Brian Thomas and Vance Nelson, "Radiocarbon in Dinosaur and Other Fossils," *Creation Research Society Quarterly* 51 (Spring 2015): 299–311.
26. Ibid.
27. Mark Hollis Armitage and Kevin Lee Anderson, "Soft Sheets of Fibrillar Bone from a Fossil of the Supraorbital Horn of the Dinosaur *Triceratops horridus*," *Acta Histochemica* 115 (July 2013): 603–8, doi:10.1016/j.acthis.2013.01.001.

**Chapter 4**

1. Marcin Wysokowski et al., "Poriferan Chitin as a Versatile Template for Extreme Biomimetics," *Polymers* 7 (February 2015): 235–65,

doi:10.3390/polym7020235.

2. Keely Glass et al., "Direct Chemical Evidence for Eumelanin Pigment from the
3. Jurassic Period," *Proceedings of the National Academy of Sciences, USA* 109 (June 2012): 10218–23, doi:10.1073/pnas.1118448109.
4. Mary Higby Schweitzer, Jennifer L. Wittmeyer, and John R. Horner, "Soft Tissue and Cellular Preservation in Vertebrate Skeletal Elements from the Cretaceous to the Present," *Proceedings of the Royal Society B* 274 (January 2006): 183–97, doi:10.1098/rspb.2006.3705.
5. M. J. Collins et al., "The Survival of Organic Matter in Bone: A Review," *Archaeometry* 44 (August 2002): 383–94.
6. Schweitzer, Wittmeyer, and Horner, "Soft Tissue and Cellular Preservation," 183–97.
7. Christina E. O'Malley, William I. Ausich, and Yu-Ping Chin, "Isolation and Characterization of the Earliest Taxon-Specific Organic Molecules (*Mississippian, Crinoidea*)," *Geology* 41 (March 2013): 347–50.
8. Matthew J. Collins et al., "A Basic Mathematical Simulation of the Chemical Degradation of Ancient Collagen," *Journal of Archeological Science* 22 (March 1995): 175–83.
9. Theagarten Lingham-Soliar and Joanna Glab, "Dehydration: A Mechanism for the Preservation of Fine Detail in Fossilised Soft Tissue of Ancient Terrestrial Animals," *Palaeogeography, Palaeoclimatology, Palaeoecology* 291 (May 2010): 481–87, doi:10.1016/j.palaeo.2010.03.019.
10. Schweitzer, Wittmeyer, and Horner, "Soft Tissue and Cellular Preservation," 183–97.
11. Collins et al., "A Basic Mathematical Simulation," 175–83.
12. Susan E. Francis, David J. Sullivan Jr., and Daniel E. Goldberg, "Hemoglobin Metabolism in the Malaria Parasite *Plasmodium falciparum*," *Annual Review of Microbiology* 51 (October 1997): 97–123, doi:10.1146/annurev.micro.51.1.97.
13. Craig A. Parish et al., "Broad-Spectrum Antimicrobial Activity of Hemoglobin," *Bioorganic and Medicinal Chemistry* 9 (February 2001): 377–82, doi:10.1016/S0968-0896(00)00263-7.
14. Mary H. Schweitzer et al., "A Role for Iron and Oxygen Chemistry in Preserving Soft Tissues, Cells and Molecules from Deep Time," *Proceedings of the Royal Society B* 281 (January 2014): doi:10.1098/rspb.2013.2741.
15. Collins et al., "A Basic Mathematical Simulation," 175–83.
16. Patricia G. Weaver et al., "Characterization of Organics Consistent with β–Chitin Preserved in the Late Eocene Cuttlefish *Mississaepia*

*mississippiensis*," *PLoS ONE* 6 (November 2011): e28195, doi:10.1371/journal.pone.0028195.

17. J. R. Nance et al., "Preserved Macroscopic Polymeric Sheets of Shell-Binding Protein in the Middle Miocene (8 to 18 Ma) Gastropod *Ecphora*," *Geochemical Perspective Letters* 1 (January 2015): 1–9, doi:10.7185/geochemlet.1501.
18. Brian Thomas, "A Review of Original Tissue in Fossils and Their Age Implications," in *Proceedings of the Seventh International Conference on Creationism* (Pittsburgh, PA: Creation Science Fellowship, 2013); Brian Thomas, "Original Biomaterials in Fossils," *Creation Research Society Quarterly* 51 (Spring 2015): 234–47.
19. Mike Buckley and Matthew James Collins, "Collagen Survival and Its Use for Species Identification in Holocene-Lower Pleistocene Bone Fragments from British Archeological and Paleontological Sites," *Antiqua* 1, no. 1 (2011): e1, doi:10.4081/antiqua.2011.e1.
20. Ibid.
21. Ibid.
22. Ibid.
23. Ibid.
24. Jack Horner and James Gorman, *How to Build a Dinosaur: The New Science of Reverse Evolution* (New York: PLUME, 2010), 73.
25. Schweitzer, Wittmeyer, and Horner, "Soft Tissue and Cellular Preservation," 183–97.
26. James D. San Antonio et al., "Dinosaur Peptides Suggest Mechanisms of Protein Survival," *PLoS ONE* 6 (June 2011): e20381, doi:10.1371/journal.pone.0020381.
27. Shuang-Yin Wang et al., "Why Collagens Best Survived in Fossils? Clues from Amino Acid Thermal Stability," *Biochemical and Biophysical Research Communications* 422 (May 2012): 5–7, doi:10.1016/j.bbrc.2012.04.122.

**Appendix A**

1. Hugh Ross, *A Matter of Days: Resolving a Creation Controversy*, 2nd exp. ed. (Covina, CA: RTB Press, 2015): 145–219.
2. Don Stoner, *A New Look at an Old Earth: Resolving the Conflict between the Bible and Science* (Eugene, OR: Harvest House, 1997): 57–116.
3. David Snoke, *A Biblical Case for an Old Earth* (Grand Rapids: Baker, 2006): 24–46.
4. Rodney Whitefield, *Genesis One and the Age of the Earth: What Does the Bible Say?* (San Jose, CA: R. Whitefield, 2011). The entire booklet

can be downloaded as a free PDF here: http://www.creationingenesis.com/Genesis_One_and_the_Age_of_the_Earth.pdf.

**Appendix B**

1. For a detailed discussion of the difficulties of cloning extinct organisms, listen to Fazale Rana and Joe Aguirre, "Can Neanderthals and Woolly Mammoths Be Resurrected from Extinction Via Cloning?," *Science New Flash*, podcast audio, May 11, 2010, http://www.reasons.org/podcasts/science-news-flash/can-neanderthals-and-woolly-mammoths-be-resurrected-from-extinction-via-cloning.
2. Jack Horner and James Gorman, *How to Build a Dinosaur: The New Science of Reverse Engineering* (New York: PLUME, 2010).
3. Bhart-Anjan S. Bhullar et al., "A Molecular Mechanism for the Origin of a Key Evolutionary Innovation, the Bird Beak and Palate, Revealed by an Integrative Approach to Major Transitions in Vertebrate History," *Evolution* 69 (July 2015): 1665–77, doi:10.1111/evo.12684; Carl Zimmer, "Reverse Engineering Birds' Beaks into Dinosaur Bones," *New York Times*, May 12, 2015, http://www.nytimes.com/2015/05/12/science/reverse-engineering-birds-beaks-into-dinosaur-bones.html?_r=0.
4. It's important to recognize that the feathered dinosaurs—interpreted as transitional intermediates—appear in the fossil record *after* the first bird appears Paleontologists refer to such an event as the temporal paradox. The out-of-sequence fossil record justifies skepticism about the evolutionary explanation for bird origins. For more, see Fazale Rana, "New Challenge to the Bird-Dinosaur Link," *Connections*, April 1, 2000, http://www.reasons.org/articles/new-challenge-to-the-bird-dinosaur-link.
5. Fazale Rana, "Archetype or Ancestor? Sir Richard Owen and the Case for Design," New Reasons to Believe (e-zine), December 2013, 4–6, http://www.reasons.org/articles/archetype-or-ancestor-sir-richard-owen-and-the-case-for-design.

**Appendix C**

1. Chris L. Organ et al., "Origin of Avian Genome Size and Structure in Non-Avian Dinosaurs," *Nature* 446 (March 2007): 180–84, doi:10.1038/nature05621.

# Index

# About the Author

**Fazale Rana** attended West Virginia State College and then Ohio University, where he earned a PhD in chemistry. His postdoctoral work was conducted at the Universities of Virginia and Georgia. He was a presidential scholar, elected into two honor societies, and won the Donald Clippinger Research Award twice at Ohio University. Dr. Rana worked for seven years as a senior scientist in product development for Procter & Gamble before joining Reasons to Believe, where he now serves as vice president of research and apologetics.

He has published articles in peer-reviewed scientific journals, delivered presentations at international scientific meetings, and addressed church and university audiences in the United States and abroad. Dr. Rana also coauthored a chapter on antimicrobial peptides for *Biological and Synthetic Membranes* in addition to contributing numerous feature articles to Christian magazines. He appears on reasons.org podcasts and gives frequent radio interviews.

He and his wife, Amy, live in Southern California.

# About Reasons to Believe

Uniquely positioned within the science-faith discussion since 1986, Reasons to Believe (RTB) communicates that science and faith are, and always will be, allies, not enemies. Distinguished for integrating science and faith respectfully and with integrity, RTB welcomes dialogue with both skeptics and believers. Addressing topics such as the origin of the universe, the origin of life, and the history and destiny of humanity, RTB's website offers a vast array of helpful resources. Through their books, blogs, podcasts, and speaking events, RTB scholars present powerful reasons from science to trust in the reliability of the Bible and the message it conveys about creation and redemption.


**For more information, contact RTB via:**

**www.reasons.org**

**818 S. Oak Park Rd.**

**Covina, CA 91724**

**(855) REASONS • (855) 732-7667**

**ministrycare@reasons.org**